ECOLOGIES OF CREATIVE MUSIC PRACTICE

Ecologies of Creative Music Practice: Mattering Music explores music as a dynamic practice embedded in contemporary ecological contexts, one that both responds to, and creates change within, the ecologies in which it is created and consumed. This highly interdisciplinary analysis includes theoretical and practical considerations – from blockchain technology and digital platform commerce to artificial intelligence and the future of work, to sustainability and political ecology – as well as contemporary philosophical paradigms, guiding its investigation through three main lenses:

- How can music work as a conceptual tool to interrogate and respond to our changing global environment?
- How have transformations in our digital environment affected how we produce, distribute and consume music?
- How does music relate to matters of political ecology and environmental change?

Within this framework, music is positioned as a starting point from which to examine a range of contexts and environments, offering new perspectives on contemporary technological and ecological discourse. *Ecologies of Creative Music Practice: Mattering Music* is a valuable text for advanced undergraduates, postgraduates, researchers and practitioners concerned with producing, performing, sharing and listening to music.

Matthew Lovett is Associate Professor in Music Innovation at the University of Gloucestershire.

'Music matters. And, as Matthew Lovett argues in this compelling book, it cannot be understood without reference to matter, whether in the form of technological tools or wider physical environments. Taking an ecomusicological approach, Lovett joins the dots between Vaughan Williams and Justin Bieber, between blockchains and physical bodies, to position music as enmeshed, embedded, entangled – and entirely interdependent with other systems in both production and consumption.'

Marcus O'Dair, *Associate Dean, Knowledge Exchange and Enterprise, University of the Arts London*

'*Ecologies of Creative Music Practice* reminds us that music, before it is anything else, is interdependent and environmental, embedded in various systems, ecologies, and material networks as it is: this has always been the case, but in a context of naturing-culturing anthropocenes, capitalocenes and novacenes this sort of immanent critical approach to music is more timely than ever. Taking an intriguing, adaptive ecomusicological approach in which recent trends in new materialist thinking are applied to and within various musical and music business contexts, *Ecologies of Creative Music Practice* brings theory and matter – the matter of theory, the theory of matter – together to illuminate the practices of music and, in turn, to use those practices to help us think differently about broader questions of technology, materialism and philosophy, and the environment. Roving across everything from video games to speculative realism, AI and blockchain to François Laruelle, and ending with a vision of music as 'ecology in motion', the book is a must read for anyone interested in creative musical practice as an assemblage or nexus of big, knotty, heavy global challenges (and vice versa). This is an intriguing book that repays close attention.'

Dr Stephen Graham, *Head of School of Arts and Humanities, Goldsmiths, University of London*

ECOLOGIES OF CREATIVE MUSIC PRACTICE

Mattering Music

Matthew Lovett

LONDON AND NEW YORK

Designed cover image: Courtesy of the Cyfarthfa Castle Museum & Art Gallery and London College of Communication (UAL)

First published 2024
by Routledge
4 Park Square, Milton Park, Abingdon, Oxon OX14 4RN

and by Routledge
605 Third Avenue, New York, NY 10158

Routledge is an imprint of the Taylor & Francis Group, an informa business

British Library Cataloguing-in-Publication Data
A catalogue record for this book is available from the British Library

Library of Congress Cataloging-in-Publication Data
Names: Lovett, Matthew, author.
Title: Ecologies of creative music practice : mattering music / Matthew Lovett.
Description: Abingdon, Oxon ; New York : Routledge, 2023. |
Includes bibliographical references and index.
Subjects: LCSH: Music--Environmental aspects. | Ecomusicology. |
Ecology--Philosophy. | Soundscapes (Music)--History and criticism.
Classification: LCC ML3799.3 .L68 2023 (print) | LCC ML3799.3 (ebook) |
DDC 780/.0304--dc23/eng/20230912
LC record available at https://lccn.loc.gov/2023030798
LC ebook record available at https://lccn.loc.gov/2023030799

ISBN: 978-1-032-12704-0 (hbk)
ISBN: 978-1-032-12703-3 (pbk)
ISBN: 978-1-003-22583-6 (ebk)

DOI: 10.4324/9781003225836

Typeset in Sabon
by Taylor & Francis Books

CONTENTS

COVER CREDIT INFORMATION

Cover image: Margaret Watts Hughes *Voice Figure No. 199*

Born in Dowlais, South Wales, Margaret (Megan) Watts Hughes (12 February 1842–29 October 1907) was a Welsh singer, scientist and philanthropist. After initial success as a singer in South Wales in her early twenties, Watts Hughes studied at the Royal Academy of Music, before continuing her musical career and developing her activities as a philanthropist in London in the 1870s. In 1885, she developed the 'Eidophone', a device that enabled her to measure the power of her voice, by translating sound into images. The Eidophone consisted of a mouthpiece connected via a tube to a bowl-like resonance chamber, over which was stretched a rubber membrane. Watts Hughes experimented by placing sand, powder, water and milk onto the membrane, which formed into patterns in response to the standing-wave resonance created by her voice. She fixed these patterns permanently on glass or ceramic plates, calling them 'Voice Figures'. A number were published in her article in *The Century Magazine* in 1891, and her book, *The Eidophone Voice Figures: Geometrical and Natural Forms Produced by Vibrations of the Human Voice*, was published in 1904.

INTRODUCTION

We humans are living in an era of rapid and profound change.

As citizens of the twenty-first century, we inhabit a world that is being transformed via environmental disruption and the continuous, seemingly exponential, evolution of technology. At the same time, our understanding of the nature and functioning of human knowledge, perception and understanding is itself undergoing a process of significant evolution, if not paradigmatic change.

In 2019, the International Commission on Stratigraphy's Anthropocene Working Group (AWG) voted in favour of adopting the term 'Anthropocene' as the 'formal chronostratigraphic unit' for our current epoch (Subcommission on Quaternary Stratigraphy, 2019). In doing so, the AWG acknowledged the presence of artificial radionuclides in the stratigraphical record; the result of nuclear weapons testing in the 1950s, a presence that offers palpable proof of how humanity has altered the stratification process of the planet. Since the work of Eugene Stoermer and Paul Crutzen first brought the concept into wider public consciousness in the early twenty-first century, questions over the meaning and implications of the Anthropocene have catalysed intense discourse amongst philosophical, creative and critical communities. In 2009, the environmental scientist Johan Rockström led the development of the planetary boundaries framework (Stockholm Resilience Centre, 2022). Entitled 'A Safe Operating Space for Humanity', the article, published in *Nature*, set the last 10,000 years of environmental stability that has characterised the Holocene, against the human-made environmental turbulence of the Anthropocene. Similarly, where the Holocene during its long duration has been marked by gradual environmental change catalysed by naturally occurring forces, leading to relatively narrow fluctuations in temperature, fresh water availability and biogeochemical flows, it appears that the hallmarks of the Anthropocene are rapid and irreversible environmental transformation. Rockström described the planetary boundaries as a framework

DOI: 10.4324/9781003225836-1

that marks 'the safe operating space for humanity' in relation to the earth's biophysical systems and processes (Rockström, 2009). He and his team identified nine of these systems, stating that, were they to cross threshold states, the consequence would be irrevocable change to the earth's environment.[1] As with 'climate change', the first of Rockström's earth systems, the word Anthropocene has been the focus of intense debate and academic scrutiny. Jason Moore, among others, refers to our current epoch as the 'Capitalocene' (Moore, 2015), suggesting that the blame for climate upheaval cannot be laid at the feet of all humanity, just a certain section of humanity; namely the capitalist, industrialist classes. Klaus Schwab, founder of the World Economic Forum, has used the phrase the 'Fourth Industrial Revolution' (Schwab, 2017), to describe our current era, suggesting that the emergence of combined physical, digital and biological technologies will impact 'all disciplines, economies and industries, and even [challenge] ideas about what it means to be human' (Schwab, 2017). Schwab's view is emphasised by the climate scientist James Lovelock, who describes our age as the 'Novacene' (2019); an era increasingly dominated by artificial intelligence and machine learning, and driven by humanity's apparent need 'to use computers to design and make themselves' (Lovelock, 2019). Given humanity's impact on the material world, it is also notable that New Materialism (Coole and Frost, 2010) has emerged as an identifiable school of philosophical thought; a project intended to re-evaluate and renegotiate ways of thinking in terms of ontologies and epistemologies, mapped to an understanding of ourselves as worlded and worlding beings.

Ecomusicological in its scope, *Ecologies of Creative Music Practice: Mattering Music* is an exploration of music which thinks through issues of creativity, production and experience in terms of these emergent issues and modes of thought; considering music in relation to our awareness of being embedded in a range of earth systems, human-made environments and conceptual frameworks. The phrase 'creative music practice' is deliberately broad. It allows us to understand how music is deeply enmeshed with these environments and frameworks, and across the book I use 'creative music practice' to refer to creative practices of making and listening to music, as well as to practices of creating contexts for music production, performance, monetization, sharing and listening. Fundamentally, this connectedness informs the book's key assertion; that thinking ecologically is to recognise the interdependent nature of music and of the many environments it interacts with. Thus, by engaging with the ecologies of creative music practice set out in this book, we can gain further insights into how music shapes, and is shaped by the many ecologies in which we create, perform, listen to, experience and share it. As Donna Haraway suggests in *Staying with the Trouble*, referencing the ethnographer Marilyn Strathern's contention that 'it matters what ideas we use to think other ideas' (Strathern, in Haraway, 2016: 34), 'it matters what thoughts think thoughts, it matters what knowledges know knowledge' (ibid.: 35). In essence, *Ecologies of Creative Music Practice* presents an account of music that challenges its readers to consider, to understand – and to hear – music within an ecological, systems-based and materialist context.

Broadly, *Ecologies of Creative Music Practice* operates across three interlinked areas of focus, so as to open out these debates for academics, music makers and the music business community alike:

- In the early years of the twenty-first century, re-articulations of realism and materialism have informed much philosophical debate; exploring ontological and epistemological issues via humankind's interconnectedness with the physical environment. Understanding and negotiating a coherent response to imperatives such as the Anthropocene and the complex interplay between natural and human-made systems is key to much contemporary theoretical discourse, and music has a vital role to play here. The book explores how music can work as a conceptual tool for interrogating and responding to our current moment of understanding ourselves, and the relationship we have with a changing global environment.
- Our understanding of the relationship between music and technology is also changing, particularly in relation to matters of value and governance, ownership and authorship, distribution and sharing, along with evermore pressing debates around sustainability and the future of work. While the global uptake of digital commerce is not in itself a twenty-first century phenomenon, the book engages with what are now common terms, such as artificial intelligence, machine learning, blockchain and platform capitalism (Srnicek, 2017), to map a set of transformations in our digital environment, and the subsequent effects on how we produce, distribute and consume music.
- In 2015, the United Nations (UN) published 'Transforming Our World: The 2030 Agenda for Sustainable Development', an agenda based on 17 sustainable development goals designed to encourage 'all countries [to] improve health and education, reduce inequality, and spur economic growth, while tackling climate change and working to preserve our oceans and forests' (United Nations, 2015). In response, the book also acknowledges that music takes its place in the world in relation to matters of political ecology and environmental change.

Ecologies of Creative Music Practice is thus built on the contention that, via an ecomusicological framework, we can position music as a starting point from which to investigate a set of theoretical, practical and technological contexts, or environments, and bring new perspectives to contemporary technological and ecological discourse. At the same time, we can also use music's embeddedness within these environments to generate a set of perspectives about making, sharing, listening to, experiencing and understanding music. To an extent, the book is a critical practice of music, one that explores how we understand music through acknowledging and engaging with a range of practices that combine to underpin, inform, facilitate and produce the practice of music in its many forms. In setting out the book, I have wanted to foreground the interdependent nature of the ideas and practices that have shaped *Ecologies of Creative Music*

Practice, and I actively draw on, present and entangle the key works that make up book's own ecology. As such, *Ecologies of Creative Music Practice* is designed to enable its readers to understand and apply a challenging set of conceptual and practical imperatives, which in themselves present an opportunity to enhance and shape our experience and understanding of music.

The remainder of this chapter has a number of functions, and serves as both an introduction to and a rationale for the subject and approach of the book. It identifies key terms and concepts, and articulates the proposed ecological convergence between music and the three areas of discourse within the book: philosophy, practices of ecology and technology. As a result, my aim in what follows is to communicate the foundations upon which *Ecologies of Creative Music Practice*'s central contentions are built.

A Beginning: Ecological Thinking and Interdependence

In the early 1970s, the environmental philosopher Arne Naess developed the concept of Deep Ecology to question and destabilise commonly held anthropocentric views about humanity's relationship with our planetary habitat, and our non-human co-inhabitants. Naess took a whole-system, egalitarian approach to reframing humanity's place in the biosphere, affirming that no singular species should have dominance, or more influence, over the global habitat than another.

Rather than simply focusing on what he saw as 'shallow ecology', a movement 'concerned with fighting against pollution and resource depletion [and whose] central objective is the health and affluence of people in the developed countries' (Naess, 1973), deep ecology was intended to articulate and emphasise relationality. In this, Naess saw 'organisms as knots in the biospherical net or field of intrinsic relations' (ibid.), and understood interdependence as much more than a simple recognition of how things interact in order to produce mutual benefit. For Naess, the relation between two things – A and B – is intrinsic to what each thing is. He suggested that 'the relation belongs to the definitions of basic constitutions of A and B, so that without the relation, A and B are no longer the same things' (ibid.).

The deep ecological position is a philosophical one. Indeed, Naess refers to it as an 'ecosophy' (ibid.). In this sense, ecosophy builds on ecological research and the interconnected and interrelated nature of the world that ecological science theorises, to establish 'norms, rules, postulates, value-priority announcements' (ibid.) that result from integrating philosophy into an ecological framework. Deep ecology takes us further to think ecologically about more than just the connections between different entities within an ecosystem; it challenges us to recognise the deep interdependence of an ecosystem's inhabitants. As a consequence, we can not only apply this approach to thinking about our planetary environment, but use it to think how, in essence, music and music economies are made of interdependent relations, which can include the relations between musicians, musical instruments and music production technologies, along with a range of digital commerce and sharing technologies.

Where Naess developed deep ecology from philosophical foundations, as another mode of ecological enquiry, political ecology draws on political economy and ecological analysis. According to Greenberg and Park, it is a development within the social sciences that explores 'the relations between human society, viewed in its bio-cultural-political complexity, and a significantly humanised nature' (Greenberg and Park,1994: 1), the latter reflecting the transformative effects of human signification which we see in the conceptual shift from 'plant' to 'fresh ingredient'. More recently, others have offered a broader definition that reflects the political impacts of both our material and discursive engagement with the environments we inhabit. For example, Little describes how the field of political ecology brings together 'human ecology's focus on the interrelations between human societies and their respective biophysical environments and political economy's analyses of the structural power relations occurring between these societies' (Little, 2007). Similarly, Robbins cites the work of the naturalist Bernard Nietschmann, whose work brought together analyses of 'social systems governing redistribution, cultural standards governing resource management, and environmental systems governing the populations of wild species' as laying the foundations of political ecology (Robbins, 2020: 44).

It is worth noting that these key vectors of political ecological research and analysis have gone on to inform much of the current discourse in relation to sustainable development, and matters of sustainability more widely. *Our Common Future*, a report produced in 1987 by the World Commission on Environment and Development (also known as the Brundtland Commission) defined the concept of 'sustainable development' as the process of 'meeting the basic needs of all and extending to all the opportunity to satisfy their aspirations for a better life' (World Commission on Environment and Development, 1987). The report drew on a political ecological framework that brought together environmental and economic perspectives, and acknowledged the tensions that existed between 'scientific accuracy and political acceptability' (Borowy, 2013). Whilst Lousley (2017) reflects on the complexities surrounding the notion of 'sustainable development', and acknowledges critique of the report's universalising approach to establishing its vision of a global future, the report laid the groundwork for what became the United Nations Department of Economic and Social Affairs' 17 Sustainable Development Goals (SDGs), a matrix of actions designed to operate as 'a shared blueprint for peace and prosperity for people and the planet, now and into the future' (United Nations, 2015). Qualifying the relationship between sustainable development and sustainability, UNESCO – the United Nations Education, Science and Cultural Organisation – describes sustainability as 'a long-term goal' in the form of a more sustainable world, while 'sustainable development' refers to the various routes and processes to achieving that goal, including 'sustainable agriculture and forestry, sustainable production and consumption, good government, research and technology transfer, education and training etc.' (UNESCO, 2021). As such, the SDGs in themselves can be seen as the UN's particular distillation of political ecology research; designed to put into practice a set of learnings that draw together the political, environmental, social and cultural vectors highlighted by the likes of Little and Robbins.

Turning to more practical, cultural and even musical matters, political ecology enables us to identify and engage with issues of environmental change in terms of everyday experience. Paul Robbins' articulation of the artificial nature of fresh food – for example vegetables or a joint of meat – is a useful image with which to illustrate the concept, and indeed the field, of political ecology; a set of practices that surround and facilitate this book's exploration of music. Indeed, learnings and strategies from political ecology fundamentally inform my approach to engaging with music as an ecology throughout this book, and of understanding music's entanglement with other ecologies. In this short statement, Robbins captures the essence of political ecological analysis:

> Freshness isn't natural. It is instead a product of capitalist transport, production and processing.
>
> *(Robbins, 2020: 223)*

Here, the idea of 'freshness' allows us to observe the multiple nested processes that this state of being contains: the transformation of a living plant or animal into the raw materials of a foodstuff, the processing, market exchange and transportation of that foodstuff onto the supermarket shelf, its purchase as a food commodity by the consumer, its refrigerated storage and possible cooking, before its final consumption as part of a meal. These are just some of the more obvious stages that we might recognise as part of the supply chain that delivers fresh food from the field to the consumer's plate. Beyond mapping these physical processes, political ecology also offers an analytical framework within which we are able to critically attend to wider contexts; for example, the transformative effect of agriculture on human relationships with the natural world. Plants become 'crops', while animals become 'livestock'. One of a number of contemporary thinkers addressing matters of environmental crisis is the philosopher Timothy Morton, who has developed a singular approach to ecological analysis and enquiry. Morton speaks of the human civilisation that has developed across the 12,000 years of the Holocene as a 'post-Mesopotamian civilisation', a view which suggests that humans cut themselves off, not from Nature as such, but from a particular view of nature: 'the establishment of a human "world", cozy, seemingly self-contained [...] bounded by wild Nature on its physical outside, and by Eden on its historical outside' (Morton, 2019: 23). Describing the manner in which this post-Mesopotamian civilisation propagates itself as 'agrilogistics', Morton refers to both the explicit processes involved in agricultural activity, along with the implicit – often hidden – logics that drive its organising principles (ibid.: 45). Their example is phosphorus, which, as an agrilogistic crop fertiliser, has caused significant climate disruption in the form of soil erosion and water pollution (Mallin and Cahoon, 2020). The 'freshness' bestowed on our food by refrigeration units is another such agrilogistic process, taking into account a number of factors, including the costs of materials and energy used in the production, transport and storage of fresh food.

The political ecology of fresh food therefore acknowledges these production factors, combines this with the ecological analysis of the environmental impacts of these production processes, and considers the transformations that result from the impact of social relations, again our post-Mesopotamian civilisation's agrilogistics of transformation from plant into crop and then onward again into foodstuff. Indeed, Richerson et al. (2001) suggest that the turn to agriculture was both a necessary condition of the onset of the Holocene, due to rapidly expanding human populations, and at the same time had been utterly impossible during the Pleistocene period which immediately predated the Holocene, due to extreme variations in weather, and the inability of subsistence systems to evolve rapidly enough to overcome the challenges caused by these variations in weather and their effect on what Richerson et al. describe as 'plant resources' (ibid.).

Invented by Thomas Edison in 1877, the Cylinder Phonograph was a device that recorded the sound of the human voice by capturing its sound vibrations as indentations on a piece of tin foil wrapped around a metal cylinder. Launched in 1878, The Edison Speaking Phonograph Company was established to commercially exploit the device, and was instantly recognised as a machine that could facilitate the 'reproduction of music' (Library of Congress, 2022). In its capacity to transform an ephemeral sound into a permanently existing physical sonic artefact, we can certainly draw parallels between sound recording technology and the transformative nature of the food processing industry. What is perhaps less immediately obvious, but no less important – certainly in terms of the subsequent exponential growth of the global music industry in the early twentieth century – is the way in which Edison's phonograph transformed and indeed invented new categories of music. It is here that we can draw a comparison between the artificial nature of 'freshness' in food, and the emergence of not only 'recorded' music, but also 'live' music in the wake of that pivotal moment of imprinting sound waves on a rotating cylinder late in 1877. Before that first instance of sound recording, all music was 'live', given that there would have been no other way for audiences to have experienced a musical performance. Thus, through his invention of the sound-recording process, not only did Edison create 'recorded music' as a new form of music, he also created its double: 'live music'. Live music was suddenly everything that recorded music was not: an ephemeral experience and an occasion for human contact that could exist outside of a system of commodity exchange. Similar to the myth of fresh food, in certain ways, live music has come to be defined as an antithesis to recorded music; we like to think that it has some kind of inherent realness, and that it is not simply the product of an industrial process. These ideas have been explored in various ways in academic literature. For example, Walter Benjamin's seminal essay 'The Work of Art in the Age of Mechanical Reproduction' (Benjamin and Underwood, 2008) explored the nature of musical authenticity and value, and Philip Auslander's exploration of 'liveness' in mediatised musical performance notably focused on notions of authenticity that resulted from the 'MTV Unplugged' series in the 1990s (2008). More recent analyses have examined the use of the auto-tune audio processing tool by artists such as T-Pain (for example, Provenzano, 2018). As we shall see, in the context of ecomusicology, this mapping of physical and conceptual environments has become increasingly visible in the context of climate change, and the interdependent nature of live and recorded music has created, and sustains, vast global industries.

As mentioned, Timothy Morton has made a notable contribution to contemporary ecological debate. Through a series of books, they have brought a particular focus to bear on the potential intersections between philosophical analysis, creative practice and ecological thinking, establishing an ecological framework that draws together things such as art, humans, non-human animals, mountains and carbonated drinks cans in a heterogeneous mesh of interrelations. In the book *The Ecological Thought*, they engage with a systems framework to think about ecology: 'If there's anything monstrous in evolution, it's the uncertainty in the system at any and every point [...] All organisms are monsters insofar as they are chimeras, made from pieces of other creatures' (Morton, 2010: 68).

One of Morton's key trains of thought throughout the book is the idea of the mesh. It is a model for thinking about the world as an ecology that avoids conventional models of 'nature' as a transcendent, holistic phenomenon that we humans and the world of organic life and inorganic matter we live in and are all part of. Morton approaches this idea with the familiar phrase that 'everything is connected' (ibid.: 33). However, they qualify this claim by suggesting that 'if everything is connected, there is less of everything. Nothing is complete in itself. Consider symbiosis' (ibid.). Morton then goes on to list sets of organisms that are made up of nested organisms, such as lichen, comprised of a fungus and a bacterium or an alga; animal and fungal cells which include mitochondria and bacteria, and the human stomach which again contains bacteria along with amoebae (ibid.: 34). In opposition to a holistic image of nature, Morton suggests that,

> the ecological thought isn't about a superorganism. Holism maintains that the whole is greater than the sum of its parts. Unlike nature, what the ecological thought is thinking isn't more than the sum of its parts .... If we want ecology, we will have to trade in Nature for something that seems more eager .... Because evolution isn't linear, the mesh isn't bigger than the sum of its parts.
>
> *(ibid.: 35)*

What we see, in embryonic form here, is Morton's growing interest in an image of ecological thinking that understands localised interactions and developments as more than just matters of straightforward system dynamics. For Morton, 'the ecological thought makes our world vaster and more insubstantial at the same time' (ibid.: 37). By this, they are suggesting that fully apprehending the natural world involves engaging with all of the world's myriad component parts and all their exponential interrelations, a process that challenges us to accept that there is no overarching framework that encompasses the natural world as a whole system. In this sense, ecological thinking is a way of acknowledging the essential contingency of the world; there is no 'Nature' that includes and explains it. The world is like it is because of the interrelations of everything in it, and things could always be other than they are. Indeed, in their reading of Darwinism, Morton challenges the myth of 'adaptationism' to show that survival is a far more contingent process, and talks of how, if a mouse were to be dropped into a climate that was colder than it was used to, its offspring would not develop thicker fur in order to survive. Instead, the descendants of the mouse might have thicker fur, simply because they were the ones more able to survive the cold.

Morton's views on systems theory provide us with an opportunity to think about how components within a great whole interact, and what we might think of the consequences of those interactions. Clearly, there is a sense in which the word 'system' conveys a sense of logical functioning; a system is a system because it is something that works, because it is something that achieves a certain end, because it has interconnected – even interdependent – parts that all operate together in order to achieve a certain outcome. This is how Donella Meadows defines a system: 'A system is an interconnected set of elements that is coherently organised in a way that achieves something' (Meadows, 2008: 11).

While any appreciable system could be seen as an aggregation of a number of interlinked subsystems, Morton takes issue with the sense of 'coherent organisation' that Meadows articulates, as it suggests that systems and superorganisms give purpose to the myriad human and non-human inhabitants of the earth. The fundamental difference is that systems theory might assume that a system's component can only be explained in terms of its capacity to operate as part of a system, whereas Morton's approach to ecological thinking – following Graham Harman's notion of Object-Oriented Ontology (Bryant, Srnicek and Harman, 2011) – steadfastly refuses to allow anything to be defined by its participation or inclusion in a greater thing (or whole). For Morton, humans and non-humans are not determined simply by being components in larger wholes, identity and substance are more complex, and are matters of interdependence, rather than hierarchical organisation. In *Humankind: Solidarity with Nonhuman People*, they use the terms 'explosive holism' and 'implosive holism' to destabilise notions of overarching systems that provide meaning and coherence for their component parts. Explosive holism is the idea that things work within a system, whereas implosive holism suggests that overall patterns are much more contingent, and subject to the interdependencies between diverse elements at a local level. These terms are designed to challenge us to create new ways of thinking about the interrelatedness of the world, and as a concept, 'explosive holism' only exists to confirm its own redundance: things do not cluster together into macrostructures which retrospectively give them meaning and purpose, and neither do things in the world become interchangeable components of the macrostructures they form, subject to and defined by the organising principles of these resultant composites. Instead, 'implosive holism' works to show how the parts of a whole continually resist attempts to confer meaning onto them by integrating them into a larger whole, or by valuing them simply as ancillary components within a whole – be that a technological, economic, political or even an environmental framework. Things resisting incorporation into a greater whole does in one sense set up a countervailing position to systems theory, but at the same time each principle reflects the other in the sense that wholes always 'subscend' their parts.

'Subscendence' is another concept Morton uses to convey the sense in which 'the whole is always less than the sum of its parts' (Morton, 2019: 102). Derived from the word transcendence, which in simple terms means going beyond, 'subscendence' is designed to allude to going down, or going below. Morton's basic image is that, if we take humankind as a whole, then humans themselves are much more than simply human. They are clearly referring to their previous work in *The Ecological Thought*, and developing a new way to describe how the components that make up wholes are themselves multiples, or aggregates, of a number of components and so on. Thinking in terms of subscendence is to see how parts of a system are not simply defined because they are components in a system, and Morton suggests that 'humankind is ontologically smaller than the humans who make it up' (ibid.: 103). This leads them to conclude that it is possible to understand the human as 'a partial object in a set of partial objects, such that it comprises an implosive whole that is less than the sum of its parts' (ibid: 104). The idea of the 'human' does not explain everything that makes us human. Subscendence, implosive holism and the ecological thought are all modes of thinking that direct us towards understanding that categories like 'human' and 'music' do not confer an ultimate meaning or teleology on their component parts. Humans and music always remain ecologies; categories that are less than the myriad material relations that constitute them, and categories that are the contingent result of particular assemblages of interdependent forces.

As I suggested above, Morton's commitment to understanding the deep and fundamental interrelations between the things in the world can be likened to an approach to apprehending the world which complements ecological thinking: interdependence. One of the underlying objectives in *Ecologies of Creative Music Practice* is to examine how music operates across multiple contexts, discourses and technological and commercial ecologies, and theories of interdependence are key to this. Interdependence is a mode of thinking that goes beyond acknowledging that the world consists of objects interrelating with each other, and instead to understand that objects in the world – and indeed the world itself – *are* interrelations; a perspective which, as we shall see, is also central to quantum theory.

The biologist Kriti Sharma notes that a number of scientific fields, including biology, physics, neuroscience and social science, have been undergoing a process of change, 'from viewing phenomena as independent to viewing them as interdependent' (Sharma, 2015: 2). Sharma's point is that there is growing consensus across scientific communities that the world only consists of interacting phenomena that are 'mutually constituted', and that things themselves only exist by virtue of their 'dependence on other things' (ibid.). Acknowledging that this lack of essence in a conventional sense does problematise established notions of what constitutes all of the apparently independently existing objects in the world, Sharma suggests that traditional categories of independence, self-sameness and continuousness – which themselves constitute phenomena such as matter, fundamental particles, physical laws, logical statements, selves, mind and consciousness – should be superseded (ibid.: 11). Sharma's articulation of interdependence therefore explodes familiar notions of independently existing, unitary and continuous categories, and pushes towards an ontology of fundamental entanglement between phenomena, such that independence is simply a resultant feature of how we experience the world, rather than a fundamental one. She goes on to qualify her position, saying that a lack of essence does not necessarily mean that things do not exist, however. Instead, for Sharma it is simply the case that essence is not a requisite for existence and formulates the term 'contingentism' to describe how things such as food and excrement have a contingent rather than an inherent existence (ibid.: 16). In her view, 'contingentism assumes neither inherent [self] sameness nor inherent difference [from other things]', and in the case of food and excrement, where she proposes that neither are possessed of properties that make them inherently different from the other, the criteria for designating food and excrement would arise 'contingently' (ibid.). Again, her point is that any sense of unitary sameness or difference in either food or excrement is entirely contingent, and develops from the interactions – discourses, practices – that such phenomena become entangled in.

As a thinker whose work has explored materialist philosophical theory in relation to quantum physics, the feminist theorist Karen Barad provides valuable perspectives that focus on points of intersection between phenomena, rather than qualities of phenomena in themselves being agents of change and creativity. Barad's work emphasises the topological nature of the world, wherein the process of mutual constitution that Sharma describes is in perpetual motion. We are unable to step outside of the topology and look at it from a position of cool, disinterested independence; we are always involved, and part of the process of constituting – and being constituted by – the world we are part of. As with Morton's concept of implosive holism, Barad's topological framework and Sharma's account of interdependence are a challenge to engage with an ontology that breaks with familiar categories of relations between things, and thinks ecologically at a fundamental level, an even deeper ecology than even Naess might have imagined. The anthropologist Arturo Escobar reflects on how, since the 1960s, ecological anthropologists have reported that in various settings across the world, human cultures do not create their social foundations on distinctions between nature/culture or human/non-human, and instead of recognising separation, many groups recognise continuity between the 'the biophysical, human, and supernatural domains' (Escobar, 2015: 2). It is with this approach in mind that *Ecologies of Creative Music Practice* approaches the subject of music, exploring how the ideas of Naess, Morton, Barad and Sharma, among others, might be put to use in the examination of music ecologies if we start to think beyond the notion of independently existing objects and phenomena, and think instead in terms of co-constitution within a topological, interdependent network of ecologies.

In the essay 'For A Realist Systems Theory', Levi Bryant is concerned with establishing two orders of meaning, one that is the product of human relations, and one that is the product of material forces. Bryant describes 'deontologisation' as the process by which we have come to understand that 'the properties we encounter in objects do not belong to the things themselves, but rather are contributed by mind' (Bryant, 2016: 72). Such a position, known as 'correlationism', after the work of the philosopher Quentin Meillassoux (2008), has been one of the most intensely scrutinised and cited contributions to contemporary philosophical discourse to have emerged from the speculative realist movement. Bryant describes how the correlationist perspective is an approach to seeing the world in terms of an individual subject's capacity to see it, that 'the subject somehow constitutes or creates its object' (ibid.: 74–5), with the result that different worlds exist for different subjects. On this reading, social relations, as a consensus of overlapping subjectivities, are a shared representation of the world that is different from what the world actually is, which for the strong correlationist position in no sense exists beyond our experience of it; in other words, there is no world 'out there', that is in-and-of itself. Using the idea of a diamond, Bryant draws on Marx's comments about the difference between intrinsic value and commodity value, the latter being a function of the 'social relations involved in producing the diamond' (ibid.: 75), qualifying this with the fact that diamonds, while they hold certain value to humans, appear to hold no value for cats.

In this sense, Bryant's ideas enable us to circle back to the earlier discussion of the artificial nature of freshness. The freshness of food is partly achieved by the action of a refrigerator keeping it at a cool temperature, a process that simultaneously contributes to global warming by releasing fluorinated gases, such as hydrofluorocarbons into the atmosphere which absorb energy and trap heat. On the question of meaning, and whether or not greenhouse gases (GHGs), require a human subject to validate their existence, and therefore their capacity to contribute to climate change, Bryant's point is that GHG emissions 'do not raise temperatures leading to the melting of polar ice because of signification but by virtue of how they prevent heat from escaping the atmosphere' (ibid.: 84). As such, while a question may remain about how we humans are able to register the reality of global warming and its effect on the environment with the same faculties we use to produce and sell diamonds, we can certainly acknowledge that heating global temperatures through GHG emissions is a different kind of behaviour from conferring a market value on crystalline allotropes of carbon. Thus, Bryant's work helps us to separate out different registers of being and meaning, and different registers of being in relation to the matter of interdependence; demonstrating how different aspects of being and reality are entangled, and how these different aspects produce a topological, constantly evolving reality. As we shall see in the next chapter, Karen Barad uses the term 'spacetimemattering' to describe and articulate this interdependent and topological nature of reality, so as to engage with one of the key critical questions of our time, that of how to create a non-contradictory ontological framework that allows for both subjective experience and objective reality.

Like Morton, Bryant also draws on Object-Oriented Ontology which has been one of the key narratives within speculative realist – and by association, new materialist – thought. While Harman could not be described as an advocate of systems thinking or of theories of interdependence, certainly not in the sense described by Sharma, some of his key ideas are relevant here. He suggests that an object – the word he uses to refer to things in the world – is 'never fully expressed in its contacts with other objects' (Harman, 2011: 282). By this he means that whenever things interact, for example humans engaging with other humans, or with inanimate objects, or indeed inanimate objects engaging with inanimate objects – neither party is completely revealed in its full reality, or exhausted by that encounter. Harman's example is the similarity between fire burning cotton and the human perception of cotton: neither of these 'encounters' with cotton fully engages with the full reality of what cotton is, since an aspect of the cotton always 'withdraws' from any encounter it is involved in. Given that neither burning nor sewing completely reveals or engages with the full reality of cotton, Harman concludes that human perception and burning are relations with cotton that are only different in degree, rather than different in kind (Bryant, Srnicek and Harman, 2011: 8). In their own way, both Barad and Morton are working to establish a similar perspective; Barad in their use of the neologism 'spacetimemattering' to foreground the entangled relations of these three elements which come together to produce spatiotemporally located moments of reality, Morton in their sense that there are no hierarchical or teleological relationships between systems and their component parts.

These reflections on the nature of interaction provide useful contexts for thinking about interdependence in relation to music. Clearly, they emphasise the importance of recognising the interrelatedness of the world, but more importantly, these ideas emphasise that we only ever apprehend the material reality of the world in fragmentary form. We don't see 'nature' as a composite whole; instead we see its component parts interacting in certain ways for certain located and specific reasons – reasons that can change – which means that the interactions can change. As with nature, so with music and ways of making, listening to, producing, sharing, distributing, performing, consuming, feeling and purchasing music. Interdependence challenges us to understand that music is no more than its interactions with the world around it; it is no more than the sum of all its interacting parts. As these parts change, so does music, and as music changes, its component parts change too.

Interdependence is increasingly being used to describe an emergent set of perspectives and practices within music, in relation to ecologies of production and distribution, but also as a means to understand the interdependent nature of composing, performing and experiencing music. For the musician and event producer Jon Davies, music and its related practices can be understood as already operating within the context of interdependence. He proposes that, 'acting interdependently ultimately means caring and affordance, and thinking interdependently interrogates the personal role of creating structures that support people, networks and beyond' (Davies, 2019). In terms of his interests in creating structures and networks that take an inclusive approach to supporting the production and distribution of music, we might assume that Davies sees interdependence in music as a development of an independent ethos for music making and distribution, an approach that has a long and celebrated history. However, he also alludes to how the concept of interdependence articulates a collapse of the 'dichotomy between artist and audience' in the way that artist–audience communities have come together to create venues for music that actively disrupt non-inclusive social structures, which foster a range of opportunities for marginalised audiences to engage with music (ibid.). He underpins this approach with the phrase 'bodies don't just occupy space but activate them' to reflect this process of developing non-exclusive music spaces (ibid.), which, in the context of Barad's, Bryant's and Harman's perspectives on discursive and material domains, suggests that music venues are the absolute embodiment of the entangled nature of discourse and matter; it is impossible to have one without the other.

In 2020, the musician Holly Herndon and technologist Mat Dryhurst launched their 'Interdependence' podcast (Bloom, 2020), a project that draws on similar principles to those of Davies; indeed, Davies references Dryhurst's ideas on how to organise the digital music industry (ibid.). Under the heading 'A $5 grad school by Holly Herndon & Mat Dryhurst, covering the people, projects and technologies shaping 21st century culture' (Interdependence, 2022), the music-oriented podcast continues to be an ongoing series of interviews conducted by Herndon and Dryhurst which explore and acknowledge the relations and interdependencies that exist between emergent technologies, platforms and infrastructures, and the production of culture. While neither Davies' nor Herndon and Dryhurst's articulation of interdependence explicitly engages with the more profound articulation of mutual constitution that we see in the work of Sharma and Morton, there is nonetheless a clear expression of the interrelated nature of technological and creative development. Across its chapters, *Ecologies of Creative Music Practice* takes up the challenge of engaging with interdependence in terms of some of its more infrastructural implications. My aim is therefore to present a set of encounters between music and contemporary philosophical theory, the future of work and artificial intelligence, digital and platform commerce that emphasise the deeper sense of co-constitution that these philosophical and biological perspectives offer. It is not simply that all things are connected, and that they rely on each other. Interdependence describes a world in which all things are connected because they are made of each other, and that because the relations between everything are in constant flux, everything in the world is in constant flux.

Mattering Music and Ecomusicology

Having established ecological thinking and interdependence as key trajectories that underpin *Ecologies of Creative Music Practice*, we can now complete this critical contextualisation of the book by turning to ecomusicology, an interdisciplinary field that explores the relationships between music, culture and the environment. As an emergent field of practice, ecomusicological research explores how, in a political ecological sense, music is shaped by, and shapes, the natural and built environments in which it is created and performed. Importantly, in terms of the framework of this book, through its exploration of the relationships between music, culture and the environment, ecomusicological practice acknowledges the interdependence of human societies and the environment, recognising that human societies are reliant on natural resources and ecosystems for their survival and well-being.

Since the late 2000s, developments in ecomusicology have significantly increased our awareness of music's environmental impact, particularly in terms of the energy consumption and materials usage associated with the production, distribution, streaming and performance of music. According to Aaron S. Allen, whose definition of ecomusicology is published in the *Grove Dictionary of American Music*, 'Ecomusicology, or ecocritical musicology, is the study of music, culture, and nature in all the complexities of those terms. Ecomusicology considers musical and sonic issues, both textual and performative, related to ecology and the natural environment' (Allen, 2013). An essential component of Allen's definition is his emphasis on how the 'eco' prefix is less a matter of *oikos* in the Ancient Greek sense of the word, relating to both ecology and economics, and more about its 'eco-critical' focus, in terms of ecological criticism, which he frames as the 'critical study of literary and other artistic products in relation to the environment' (Allen and Dawe, 2016). Thus, ecomusicology is defined as a means by which we can consider some of the practical consequences of understanding how music is embedded in a range of environmental systems; be they earth systems, conceptual frameworks or technological platforms and digital markets.

There are numerous ways of thinking about what an ecomusicological perspective might be. Allen and Dawe list an extensive set of research trajectories:

> ecoethnographic justice, the retention of biodiversity for future generations via sustainable musical instrument making, the impacts of protest in, by,

> and / or on song, the co-survival of indigenous cultures and ecologies, the reinterpretation of canonical and non-canonical figures, the mutual interests of music psychology and ecological psychology, the contributions of non-Western cultures to ecomusicological and broader Western thought, unethical exploitation of music and natural resources, critical theory, ecocritical, and postcolonial approaches, the acoustic commons, and our ecological imaginations.
>
> *(Allen and Dawe, 2016: 6)*

Because of this breadth, ecomusicological research has done much to articulate how music is deeply involved in our current phase-change of human–earth relations; not only mapping how music as a creative practice can sound ecological enquiry, but also how ecomusicology, as an ecocritical practice, can enable us to engage with a set of ecological perspectives (in other words, ecology as method, rather than ecology as objective). Here, the work of animal behavioural ecologist Alice Boyle and ethnomusicologist Ellen Waterman provides a valuable indication of how, by employing an ecomusicological framework, they were able to establish how 'a musical performance is ecologically performative' (Boyle and Waterman, 2016: 31). Their aim was to examine how a performance affects, and is affected by, 'the cultural, social, and physical environment in which it is manifest' (ibid.), although they defined their approach as definitively 'not environmentalist' (ibid.: 29). Instead, the object of their research was a group of jazz musicians who, during a concert of improvised music clearly responded both to each other and to the audience as their improvisations progressed. In order to engage with the dynamics of the performance in a meaningful way, Boyle and Waterman developed an analytical framework that took into consideration the impact of space, place and time on the 'relations, causes and effects' that constitute a musical performance (ibid.: 36). By mapping the dynamics of environments and ecosystems beyond the specific contexts of earth systems analysis, in a broader sense, their work is a proposal for how an ecological analysis can operate; one that recognises the myriad relations and non-human forces that constitute the production of music.

In 'The Cost of Music', Brennan and Devine explore the cost of music in two senses of the word: the cost of purchasing music (either as a product or via streaming services) and the cost to the environment of producing and distributing music (again, as both a physical product and via streaming services). Their analysis suggests that, while the cost to the consumer of accessing and listening to recorded music has never been lower, the cost of listening to recorded music, in terms of its impact on the environment, is likely to be at an all-time high (Brennan and Devine, 2020: 59–60). Across a series of research projects, Devine and Brennan engage with music's impact on the material world in terms of its matrix of supply chains, including musical equipment and instrument production, energy consumption, transport costs and event costs. While this work demonstrates increased awareness of music's environmental impact, it is also an awareness-raising project that seeks to embed music evermore firmly within the sustainable development discourse.

In his own analysis of the UK live music sector, Brennan suggests that it accounts for approximately 75 per cent of the greenhouse gas emissions generated by the UK music industries as a whole (Brennan, 2020), given its deep integration with a range of national and international transport infrastructures (air, sea and road travel). It would seem, though, that within the live music festival sector, work is happening to establish sustainable models for supply management and waste disposal in the context of circular economy models. For example, in 2019, Live Nation Entertainment – the umbrella company comprising Ticketmaster, Live Nation Concerts and Festival Republic among others – launched its 'Green Nation' environmental sustainability charter. Framing Live Nation as 'stewards of the environment', the charter pledges to 'reduce the environmental impact of our venues and festivals to ensure we are being responsible global citizens' (Live Nation, 2019). Updating the pledge in 2021 with the launch of the Green Nation Touring Program, Live Nation has sought to expand its approach beyond the production of festivals and events, and to implement sustainable practices across all aspects of the music touring process, via a focus on planning, production, sourcing and community (Live Nation, 2021). What these two programmes demonstrate is a recognition on the part of a major corporation in the live music sector that the production of live events draws on a varied set of supply chains, procurement processes and human behaviours. In particular, the 2021 update shows the need for focused policies that go beyond scope 1 and 2 emissions inventories (an organisation's own greenhouse gas emissions, and emissions 'associated with the purchase of electricity, steam, heat, or cooling'), and emphasises strategies for engaging with scope 3 emissions which 'result [from] activities from assets not owned or controlled by the reporting organisation' (United States Environmental Protection Agency, 2023). In addition, wider movements within the music industry suggest an increased awareness of, and willingness to engage with the UN's 2030 Agenda. For example, in 2017, Keychange launched its '50/50 by 2022' campaign with the intention of encouraging music festivals to commit to ensuring an equal gender balance for live performers, and in 2019 a number of prominent musical artists began to question the sustainability of international touring in terms of environmental impact, including Coldplay and Massive Attack (Pidd, 2019 and Petrusich, 2020). By 2022, 600 global festivals and music organisations had committed to providing 'more opportunities for women and gender expansive creators and professionals throughout the music industry' (Keychange, 2022).

Framing his project as 'a political ecology of the evolving relationship between popular music and sound technology since 1900' (Devine, 2015: 367–8), Devine's work interrogates the environmental impact of music in relation to its economic and cultural value. He writes that 'the entire history of recorded music has been defined by material limitations, resource scarcity and energy consumption' (Devine, 2019: 133), and alludes to the misconception that in our current era, where streaming is the dominant mode of music consumption, the music industry's carbon footprint has diminished, due to the immaterial nature of digital music. On the contrary, by referencing greenhouse gas emissions as a measure of the energy intensity of the digital music sector, Devine highlights that the environmental impact of music 'is now greater than at any time during recorded music's previous eras' (ibid.: 134).

Brennan's and Devine's insights into the environmental consequences of music streaming, merchandise production and distribution, and the live events sector draw on established approaches in ecomusicology and the political ecology of music. In *Ecomusicology: Rock, Folk and the Environment*, the ethnomusicologist Mark Pedelty explored 'the conundrum of making music sustainably' (Pedelty, 2012: 5), suggesting that internationally touring rock musicians 'act out our conundrums in public' (ibid.: 26). Pedelty's contention was that the composition, production, reception, distribution and consumption of music are all ecological matters, both in the sense of music's fundamental material connection to its environmental context, but also in how music is continually part of, as well as the result of, interacting systems. The conundrums that Pedelty alludes to are the contradictions that can arise when the material processes involved in the production of music and live events are understood to be very much at odds with the environmental and ethical messages that musicians and events promoters might want to convey.

Two of Pedelty's examples illustrate these conundrums in action: the rock band Soundgarden's song 'Hands All Over', written as a commentary and lament for environmental degradation, and the 24-hour continuous Live Earth concert, which was intended to 'transform music fans into [environmental] activists' (ibid.: 24). Alongside highlighting such complexities, Pedelty also seeks to engage with how musicians have actively attempted to engage with sustainability across their practice. In the context of manufacture and materials use, he describes how the musician Craig Minowa, of the band Cloud Cult, founded the record label Earthology as a means of orienting music production and manufacturing processes towards more sustainable and environmentally friendly practices (ibid.: 123). While these are historical examples, they still function as useful illustrations of the challenges facing environmentally engaged creative practice and commercial activity, in that it can be very hard to escape the conditions of production. In 2021, the band Coldplay announced a return to global touring with a comprehensive plan to reduce the carbon footprint of their tour (BBC, 2021). As with Pedelty's examples, Coldplay's plan was not without its own complications, as alluded to by their interviewer: 'When rock stars speak about the environment, there are always cries of hypocrisy, especially when private jets are being used' (BBC, 2021). In a self-reflexive response that demonstrates not only how well established the debates now are around sustainability and energy use, but also the extent to which Pedelty's conundrums are also part of those debates, Coldplay lead singer, Chris Martin replied, 'We're trying our best and we haven't got it perfect, absolutely. And the people that give us backlash for that kind of thing, for flying ... they're right. So we don't have any argument against that' (BBC, 2021).

Pedelty's ecomusicological analysis engages with a number of pop, rock and folk tunes, examining their relation to their composers' intentions and the responses of their listeners. He reflects on how – as a piece of political geography – Woody Guthrie's 'This Land Is Your Land' communicates its protagonists' sense of self as something produced by their relation to the land – as inhabitants and workers of the land – rather than as citizens of an American polis, as such (Pedelty, 2012: 51). Neil Young's 'After the Goldrush', 'Natural Beauty' and 'Be the Rain', along with John Denver's 'Take Me Home, Country Roads' and 'Thank God I'm a Country Boy' are all framed in terms of their capacity to communicate environmental messages which in turn enable listeners to create a connection with – and discuss – issues relating to sustainability (ibid.: 73). In this way, popular music song forms become a vehicle through which musicians' own environmental concerns are shared, and at the same time, Pedelty contends that listeners' awareness is built up, as is their capacity to formulate discourse around environmental themes. His work establishes a number of important vectors that mark out music's complex relationship with the environment. Alongside his emphasis on music's material impact on earth systems, Pedelty's nuanced analysis of how a range of music workers – including producers, performers and promoters, and music consumers – are caught up in a socio-economic and ecological framework that can problematise the environmentally progressive positions frequently taken up by musicians and music professionals, provides us with a valuable foundation for much of the work in *Ecologies of Creative Music Practice*. As we progress through the various analyses of music in relation to blockchain, catalogue funds, artificial intelligence and the creator economy, Pedelty's work is a salient reminder that all these techno-economic environments operate in relation to earth systems and are unavoidably entwined with matters of environmental impact and sustainable development.

Ecomusicology and Acoustic Ecology

Alongside the work of ecomusicologists such as Allen, Pedelty, Brennan and Devine who have brought a focus on the environmental impacts and discursive complexities associated with music production, performance and distribution, the use of music and sound as tools for ecological analysis also has a well-established history. R. Murray Schafer and Barry Truax have long taken an ecomusicological approach to thinking about sound and environment, in their explorations of acoustic ecology, communication and notions of soundscape. *The Soundscape: Our Sonic Environment and the Tuning of the World* (Schafer, 1977) and *Acoustic Communication* (Truax, 1984) established precedents for thinking about the relationship between human-produced sound and the natural environment, and reciprocally, naturally occurring sound and its impact on human conceptions of music, particularly in terms of sound aesthetics and musical form. As the title of his book suggests, Schafer's work foregrounded sound as a fundamental measure of the health of natural habitats, constructed environments, human societies and cultures. Within his conception of soundscapes, Schafer recognised that such environments could comprise music, radio and the sounds of industrial society, along with the huge diversity of naturally occurring sounds and noises. By framing listening as a means to reflect on the impacts of industrial development and acknowledging the impact of noise pollution on physical and mental health, Schafer proposed that 'the general acoustic environment of a society can be read as an indicator of social conditions which produce it' (Schafer, 1977: 7). For Schafer, this led him to conclude that our current 'inhuman' environments, which overwhelm the human voice and physically endanger and psychologically debilitate the human ear, should be rebalanced using 'acoustic design' to create social and physical environments that are more in tune with the sounds of the natural world, and therefore more conducive to human health and well-being (ibid.: 7). Schafer concludes the book by proposing that 'we need to regain quietude [...] when there is no sound, hearing is most alert [...] still the noise of the mind: that is the first task – then everything else will follow in time' (ibid.: 259). Such a poetic climax to a book about listening to the sounds of the world and to a diversity of environments, but which ultimately seems to think about sound largely in relation to human hearing and vocal expression, is not entirely free from contradiction. The conundrum here is that Schafer's positioning of the human voice and human hearing as acoustic design modules presents us with something of an acoustic Vitruvian Man. Where Leonardo's visual characterisation of the proportions of the human form was intended as an analogue for the functioning of the entire universe, Schafer's modules suggest that human listening is the measure of all sound. As we have seen in the core principles of deep ecology, our human-centric viewpoint is only one of a multitude of perspectives that populate the planet, an issue we shall further explore in the next chapter. Nonetheless, despite the fact that Schafer does not engage with non-human contexts for listening in a way that a specifically deep ecological analysis might do, his work does provide us with valuable materials for thinking about sound within an ecological framework, particularly in terms of the interrelations between sound, music and human culture.

Truax developed what he called 'a communicational model' of audio-ecological thinking; a holistic approach to understanding how sound enmeshed and 'defined the relationship of the individual, the community, and ultimately a culture, to the environment and those within it' (Truax, 1984: 3). This wider view of sound production and perception acknowledges how the movement of sound operates across two modes, which Truax defines as energy transfer and acoustic communication. He suggests that, whereas hearing 'is the processing of acoustic energy in the form of sound waves and vibration', listening is more aligned with communication processes (ibid.: 9). In this, Truax highlights how listening involves processing sonic information in order to create meaning in the human brain, a two-stage process that bears resemblance to Levi Bryant's ordering of meaning, where material processes create environments that, as a result of human perception, are interpreted in various ways. As with Schafer's conception of the soundscape, Truax's communication model appears to give primacy to human cognition as the key component of an acoustic ecological framework. Unlike Bryant's materialist-inflected approach, which would allow for relations between material flows of sound and material cognitive processes that are different by degree, Truax's model seems to be based on a difference in kind between hearing and listening, where somatic flows of sound-energy-hearing contrast with mental processes of communication-context-listening that reflect a Cartesian separation of body and mind. Again, questions as to the material relations between physical systems and mental processes have been at the heart of the new materialism and speculative realism projects, and we shall turn our attention to these issues in the next chapter.

Putting to one side questions regarding the materiality of mental processes, in terms of ecological listening, the work of the musician and naturalist Bernie Krause is a significant development that built on Schafer's and Truax's research. One of Krause's key propositions was that 'natural soundscapes are the voices of whole ecological systems' (Krause, 2012: 27), an idea that positions soundscapes as sonic manifestations of an ecosystem, and therefore indicators of the health, or lack of health, of these complex environments. He determined three types of sound: 'geophony (from wind, rain, earthquakes and other natural, non-living sources); biophony (from non-human animals); and anthrophony (from humans and their machines)' (Krause, in Hoffman, 2012). Redolent of current perspectives on how the Anthropocene marks a new phase in the earth's history characterised by the pervasive and self-reinforcing impact of humans on the natural world, Krause's distinctions set the human-made anthrophonic soundscape of the capitalist-industrial Anthropocene against established, naturally occurring sounds. He suggests that a balance of organic and nonbiological elements within a habitat can provide information on the health of that habitat (Krause, 2012: 68), and he recounts his experiences of creating recordings of 'bioacoustic vitality' (ibid.: 70), discussing the capacity of sound to convey rich and detailed information about the contents and any changes occurring within a recorded environment. As such, he contends that sound recording can provide an impression of 'space and depth' that reveals 'multilayered ongoing stories that visual media alone can never hope to capture' (ibid.: 72); a compelling claim for the power of sound to communicate complex information and narratives. At the very least, we might allow Krause the distinction that visual and sonic media communicate in different ways, and display different capacities that make each more or less suitable for different tasks. In this regard, a key point of focus for Krause was his interest in acoustic territories, which he apprehended and investigated via 'an analysis of biophonic expression' (ibid.: 100). He came to understand that listening to a natural soundscape enables us to define 'a biome's borders' (ibid.: 101), and his soundscape recordings showed that 'each cohesive habitus expresses itself through its own special niche composition – its unique voice' (ibid.: 102). Thus, Krause presents us with a thoroughgoing view of how sound can define the identity of an environment, and how recording and listening to sound can enable us to diagnose the health and changing nature of that environment.

Less a matter of communication as such, at least in the sense that Truax's work suggests, Krause – while similarly foregrounding the human ear as the point of measurement and determination of identity and health of a given ecosystem – enables us to understand that human sound production is but one mode of sound, albeit an increasingly dominant mode. Although his distinctions between the 'natural' biophonic and geophonic soundscapes and the artificial, synthetic anthrophonic soundscape may prove to be problematic, particularly with regard to Morton's conception of ecological thought, Krause's work does enable us to recognise that soundscapes are inter- and intra-dependent. This is to say that, for Krause, sonic interdependence is a recognition that inorganic, biological and human-made ecosystems and soundscapes exist in relation to each other, and are defined by their likeness, difference and impact on each other, a deep ecological position reflected in his lament that 'anthrophony is getting harder to escape' (Krause, in Hoffman, 2012). With 'intradependence', Krause was referring to the changing relations within a soundscape. These relations can provide an indication of an ecosystem's health, and to illustrate Krause's ideas we can turn to the work of the ecologist and journalist, George Monbiot. In his writings about rewilding, and what he sees as the desertification of upland landscapes such as the Cambrian mountains in West Wales, Monbiot reports of the dearth of birdlife in these areas that would have once been teeming with birds as part of a vibrant animal ecosystem. Describing a visit he made to a farm in the Cambrian mountains to investigate sheep grazing, Monbiot wrote: 'as usual in the Cambrians, no birds called and nothing rustled in the grass' (Monbiot, 2014: 173). His argument, which connects to the wider project of rewilding to rebalance natural ecosystems and secure biological sustainability, suggests that the lack of birdsong is evidence of lost habitats, a situation which in itself speaks to environmental insecurity and fragility. Writing of his experience of listening to his surroundings during a walk on what he refers to as the 'sheepwrecked hills' of mid-Wales (a reference to Monbiot's view that over-grazing by sheep is the cause of upland desertification), the ecologist Mick Green also conveyed his dismay at the lack of animal sound in the hills of mid-Wales: 'there are no bees buzzing, grasshoppers chirruping or butterflies flying and the silence is eerie' (Green, 2016). Both Green and Monbiot firmly lay the blame for this environmental degradation on longstanding agricultural practices that have disrupted even longer-established natural ecosystems. As suggested by Schafer and Krause, it is the human-made sounds that invade the natural soundscapes. However, while Schafer might encourage us to embrace silence as a way of meditatively tuning the world, silence is the human-made sound that indicates the disruption caused to the nonhuman habitats of the Welsh uplands. For ecologists like Monbiot and Green, silence is not natural. It might be curious to think of silence as a synthetic sound, but in contrast to Green's experience of walking in the Pyrenees, where 'bees were buzzing and several species of grasshopper giving a constant background soundtrack' (Green, 2016), it becomes clear that the boundaries between so-called natural and human-made sounds can be complex.

Listening to the world, and creating music that evokes or integrates its soundscapes has long been a focus for composers. In *Singing in the Wilderness: Music and Ecology in the Twentieth Century*, the musicologist Wilfrid Mellers builds a perspective around a variety of musical works that frames their composers' creative engagement with the natural world. Mellers is interested in the way in which certain works demonstrate a musical encounter with the non-human that is more than simply sonic. The Brazilian composer Heitor Villa-Lobos' *Rudepoema* (literally 'savage poem') is a 20-minute piece for solo piano that requires its performer to engage with a variety of syncopated dance rhythms juxtaposed against chromatic and non-diatonic virtuosic flourishes that scale across the full range of the keyboard (Villa-Lobos, 2015). Mellers suggests that Villa-Lobos was able to create a piece that signified the wildness of the rainforest, the 'polymetrics of fiesta', '"civilised" Portuguese colonists valsing in the salons, and a Chicago-style barrelhouse pianist in a club' (Mellers, 2001: 85). There is much in the piece that demonstrates Villa-Lobos' capacity to use musical gesture to evoke a series of non-musical scenes. However, what is notable about Mellers' analysis of *Rudepoema* is that for him, its evocations of different environments is only part of what qualifies it as an ecological piece. Mellers sees that Villa-Lobos' eco-musical vision is *Rudepoema*'s capacity to conjure in sound a multitude of co-existing Brazils, and that the different tableaux that comprise the piece demonstrate Brazil's overall instability. In this sense, it is a *subscendant* piece of music, in the way that Morton insists that a whole system can never confer a fixed and stable meaning on its various component parts. The Brazil that we hear in *Rudepoema* is unstable, and very obviously the result of a number of disparate elements, all in fluctuating relationships with each other, as each one vies for dominance. Mellers attributes Villa-Lobos' capacity to evoke such a 'world in flux' to the composer's own musical predilections for Italian opera, romantic piano music by the likes of Chopin, Liszt and Rachmaninoff, and music of the 'Russian nationalists', such as Borodin, Tchaikovsky and Mussorgsky (ibid.: 84). Part of Mellers' skill in creating an analytical context around his own version of 'music and ecology in the twentieth century' was his ability to target pieces that gave voice to his very particular conceptualisation of ecology. In this sense, his response to the second and third movements of Villa-Lobos' *Suite for Voice and Violin* is again striking:

> The second [movement] mingles fragmented words and wordless vocalise as it gravitates from human articulacy toward the unnotatable hubbub of nature; in the third song the girl seems to become a forest creature, giving vent to avian squawks and screeches and beastly growls and grunts.
>
> *(ibid.: 86)*

While this reading of the piece lacks the broader ecological vision that hears *Rudepoema* giving voice to the rainforest, the Brazilian festival, the colonists' salon and the American bar, the sense of transformation that Mellers describes is nonetheless powerful. Here, the idea that the sounds made by a human voice do not simply evoke or symbolise that natural world, but to a degree become animal vocalisations, gives greater scope to how we might understand ecomusicology in the context of composition. The difference between music's capacity to invoke, and its capacity to inhabit non-musical environments is therefore key to understanding this aspect of ecomusicological analysis. Mellers' perspectives on the transformation of the female voice in the *Suite for Voice and Violin* into something non-human is deeply redolent of both Naess' and Morton's views on the interdependent nature things, where everything in the world is made of everything else in the world. For Mellers, music is the means by which composers are able to articulate, and listeners are able to discern, the deep relations between humans and non-humans, where the human voice is able to inhabit the liminal acoustic space between human and non-human expression.

In his examination of Debussy's *Pelléas et Mélisande*, Mellers establishes its creative context, positioning the opera at 'the end of the cycle of European humanism' (Mellers, 2001: 22). For Mellers, this cultural shift is important, as on his account, it informs Debussy's use of static harmonies and isolated leitmotifs throughout the opera to signify a 'will-lessness' (ibid.: 23) and a retreat to 'the inner life' (ibid.). Mellers sees Debussy's approach in the opera as a reaction to those of his predecessors, for example Wagner and Schoenberg, whose work could often be understood as a dramatisation of the 'Western pain of consciousness' (ibid.). Mellers proposes that *Pelléas et Mélisande* presents human consciousness as 'flux [...] as existence without duration' (ibid.), which places Debussy's work in a more radical line of development as regards its framing of the relationship between humans and the natural world. According to Mellers, it would seem that, for Debussy, human consciousness could be understood less as a striving to overcome nature, and more as a will to be subsumed within, and as part of the natural world. Certainly, in comparison with the jarring, chromatic aesthetics of Wagner and the strident, expressionistic stylings of Schoenberg, which more obviously evoke the inner turmoil and heroic, personal struggle associated with High Romanticism, the languidity of Debussy's music in *Pelléas et Mélisande* decentres the Romantic hero. First performed in 1902, Mellers hears in the opera an early indication of the ecological thinking that, as we have seen, later found expression in the work of Arne Naess. This is not to say that *Pelléas et Mélisande* was specifically attuned to conceptions of deep ecology *avant la lettre*. However, Mellers' interpretations draw our attention to how Debussy is able to communicate the idea that, after the intensely humanistic compositional stylings of the nineteenth century, which can be seen as the zenith of European Enlightenment thinking, as we entered the twentieth century, music was evolving, becoming increasingly capable of articulating a growing awareness of humanity's relation to non-human ecologies.

In *The Place Where You Go to Listen: In Search of an Ecology of Music*, the composer John Luther Adams sets out his credo, describing how ecological thinking has informed his approach to creating music, in a manner that is redolent of the interdependence and systems-based perspectives we have already encountered in this chapter. It is worth quoting Adams at length to gain a sense of the breadth of his vision for his creative practice as a musician:

> An ecosystem is a network of patterns, a complex multiplicity of elements that function together as a whole. I conceive of music in a similar way [...] The central truth of ecology is that everything in this world is connected to everything else [...] As a composer, I believe that music can contribute to the awakening of our ecological understanding. By deepening our awareness of our connections to the earth, music can provide a sounding model for the renewal of human consciousness and culture. Over the years this belief has led me from music inspired by the songs of birds, to landscape paintings with tones, to elemental noise and beyond, in search of an ecology of music.
>
> *(Adams, 2009: 1)*

There are two important ideas here that inform Adams' approach to making music. First, given his interest in systems thinking, he understands that music by its nature is part of a wider ecology of both material and immaterial earth systems. Second, he proposes that music's ecological nature – in other words its embeddedness and interdependence within a web of ecosystems – endows it with the capacity to speak to its listeners of its own ecological nature and of the wider ecology of which it is a part. In this regard, Adams' position resembles the ecomusicological listening practices of Schafer, Truax and Krause, but where their work draws on the material connections between listener and environment, building up knowledge of various soundscapes so as to develop understanding about the connections between humans and our biological and non-biological environments, Adams' work is more conceptually rooted. For him, it is by recognising and understanding music's ecological nature that we come to understand the ecological nature of the world. In Bernd Herzogenrath's view, 'Adams does not represent nature through music. He creates tonal territories that resonate with nature' (Herzogenrath, 2012a: 1). Indeed, Adams charts his own progression as a composer of moving beyond the task of evoking particular places and experiences, towards his later practice of attempting to move beyond the particular, and instead engage listeners with a more pervasive sense of being within an ecological environment:

> For more than a decade I composed musical landscapes. My experience in wild places inspired choral and orchestral works [...] My music became less pictorial as I aspired to evoke the experience, the feeling of being in a place, without direct reference to a particular landscape.
>
> *(Adams, 2009: 1–2)*

Communicating a sense of what being somewhere feels like is not only a challenge for a composer, but a complex perspective to apprehend. With this sentiment, Adams is expressing a sense that the experience of music can take us beyond representations or connections to recognisable places or experiences, and that music can be used to communicate, evoke and enable the listener to reflect on what *being* feels like. In Adams' music, being is an ecological matter; an experience of understanding or feeling oneself existing as an interdependent part of a wider, heterogeneous whole.

In the essay 'The Weather of Music: Sounding Nature in the Twentieth and Twenty-first Centuries', Herzogenrath writes that

> whereas the composers of the eighteenth and nineteenth centuries were mainly interested in the representation of the subjective effects of weather phenomena, the avant-garde more and more focuses on the reproduction of the processes and dynamics of the weather as a system on the edge of chaos.
>
> *(Herzogenrath, 2012b: 219)*

Herzogenrath's perspectives are compelling here, as they reflect Boyle and Waterman's formulation of a framework for ecological analysis that is not in itself designed to engage directly with earth systems. Herzogenrath's contention that Adams' music – as evidenced in the piece *The Place Where You Go to Listen* – is less a representation of weather forms, in other words, the music is not intended to sound like the weather, and instead is 'coextensive with the environment' (ibid.: 230), effectively conveys Adams' interest in ecological dynamics, rather than mimicking weather sounds as aesthetic effect.

Adams has explored this approach in a number of his works, notably in the piece *Inuksuit*. The Inuit word Inuksuit is the name given to the stone structures that mark the northernmost boundaries of the Canadian arctic; the end point of human civilisation and the beginning of the uninhabitable polar ice. Adams named the piece after these objects that stand as 'a point of demarcation between people and wilderness', writes the percussionist Steven Schick (2012: 98). Schick, one of the percussionists who played at the world premiere of *Inuksuit* at the Banff Centre for the Arts in 2009, describes it as,

> The truly rare music of the second person – an exquisite, hortatory state where a collusion of energies between humans and the natural world creates a music that no single person can make in a space that no single person can hear.
>
> *(ibid.: 105)*

In a compelling passage, David Shimoni articulates Adams' compositional challenge in *Inuksuit*:

> Instead of making music from nature, in which nature is treated as a resource, can we make music with nature, in such a way that both humans

> (composer, performers, listeners) and the rest of the natural world retain at least a sense of autonomy and creativity in the process?
>
> *(Shimoni, 2012: 237)*

By delivering on this ambition and giving voice to a multiplicity of human- and non-human-made sonic materials, and creating *Inuksuit* as a piece to be performed outdoors, Shimoni suggests that Adams was indeed able to harness the natural world as more than something to be simply heard. As a result, Shimoni identifies the piece as 'ecocentric' music (ibid.: 266), where Adams was able to move beyond familiar musical techniques of imitating and representing nature with music, beyond using natural sounds as a component of music, and even, perhaps most importantly, and beyond self-expression and an anthropocentric approach to making music (ibid.: 237). This sense of superseding an anthropocentric viewpoint in order to engage with non-human perspectives – which moves us beyond John Cage's familiar dictum that music is 'sound heard' – brings a distinctly nonhuman perspective to bear on Truax's notion of acoustic communication. In the same way that Truax proposes that it is essential for us to understand that listening to sound is more than just a matter of energy transfer, *Inuksuit* is an attempt to make music with what Morton has described as 'nonhuman people' (Morton, 2019). In other words, *Inuksuit* is Adams' attempt to think about listening rather than just hearing, and to imagine what acoustic communication might mean to nonhuman listeners. It is a fusion of material-discursive conceptions of sound, which can be likened to the perspectives we have encountered in Naess', Morton's and Barad's work. *Inuksuit* is a piece in which systems do not fully account for their component parts, they are simply produced by them. Moreover, it is music that challenges us to understand that humans are not the ultimate recipients and decoders of sound. Just as Harman insists that neither fire nor humans can fully grasp the meaning of cotton, Adams' work in *Inuksuit* is an expression of the idea that in a world full of non-humans, we humans are only one type of listener.

In this overview of ecomusicological research, we have seen how Pedelty and Devine examine the numerous tangible impacts on the environment that result from the materials and energy use involved in the production, distribution and performance of music. Schafer, Truax and Krause have enabled us to hear sound and music as part of a set of physical as well as discursive environments, where the existence of sound allows us to register and record certain features of the physical environment – at least everything that either makes an audible sound, or emits something that can be rendered as sound – and enables us to speculate about the nature and function of sound. As Truax suggests, acoustic communication expresses the sense that naturally occurring sound, language and music all exist in a continuum, which for Mellers, manifests in Villa-Lobos' and Debussy's ability to express through music, this deep interrelation of human and non-human worlds, and the fragile, contingent nature of ecosystems. Using Naess' concept of 'deep ecology', we can now harness these ecomusicological approaches and begin to think in terms of a 'deep ecomusicology'. Here, interdependence enables us to recognise that things are as they are because of their interaction with other things, that things in the world only are what they are because of the multiple interactions they are having with everything else, all of the time. In this way, a deep ecomusicological approach which draws on the perspectives on interdependence espoused by Naess, Morton, Barad and Sharma is a way of thinking about music that recognises a fundamental relation between music and an array of earth systems and technological, economic, philosophical and cultural environments, and understands how changes in any one of these interrelations have the capacity to recalibrate and change the other. This is the key trajectory that runs throughout *Ecologies of Creative Music Practice*, and the following section describes how this approach takes form in a series of music-related enquiries across the book's chapters.

Structure of the Book

Systems of thought, such as New Materialism, Agential Realism, Speculative Realism, Object Oriented Ontology, Deep Ecology and Systems Thinking, which foreground the embedded nature of human experience within a material environment, all offer opportunities for framing musical creativity and experience. Chapter 1 explores a range of critical and philosophical perspectives in order to establish a foundation for the book's overall approach to engaging with music. The focus of this chapter will be to articulate how contemporary thinking systems are addressing the relationship between human thought and the multiple environments that we live and work in. The chapter presents an analysis of music in relation to a set of thinking systems, drawing on the work of Karen Barad, Quentin Meillassoux, Timothy Morton, Donna Haraway and François Laruelle, amongst others. It also responds to recent critiques of the new materialism project in terms of its relation to political, philosophical, and most importantly, musical, discourse. The chapter concludes by conceiving of a speculative materialist ecology that will allow us to fully engage with music as an ecology that is always evolving, and as a consequentially non-human force that has the capacity to produce real change in the world.

Chapter 2 generates a set of perspectives that offer insight into how ecological thinking and theories of interdependence can be identified and explored in relation to musical creativity, in terms of composition, improvisation, performance and listening. Having laid the groundwork for thinking about creativity and experience within a set of contemporary philosophical and critical frameworks, this chapter engages directly with these key modes of musical practice. Making and listening to music are material processes, that involve working with and manipulating physical bodies, generating and responding to disturbances in the physical world, and recognising the physical foundations of mental and emotional experiences within the human body. From the seemingly simple act of depressing keys on a piano keyboard in order to create chords and melodies, to the more complex processes of designing sound within immersive audio environments, this chapter maps out an approach to understanding what is happening when we create and listen to music, through the lens of interdependent material interactions, or what the philosopher and theorist Karen Barad refers to as 'intra-actions'.

The following three chapters explore music in relation to technology from three vantage points, examining how music as a creative practice is informed and moulded by an evolving technological environment, in particular the interdependent relationship between music, production and consumption and a range of emergent technologies.

Chapter 3 presents an analysis of music in relation to emergent models of intellectual property rights, revenue management and digital governance. The chapter explores the part that recorded music's current format as an intangible data asset is playing in catalysing change in music production and consumption habits, and I focus on how rights and revenue patterns are evolving as music becomes increasingly entangled with song funds and decentralised computer architectures, such as blockchain-based streaming platforms and NFTs (non-fungible tokens). The chapter also explores how value systems, market dynamics, financial products and indeed money itself in the early twenty-first century are evolving, and how, within a systems framework, these changes are connected to broader social, political and ethical patterns. By investigating how emergent technological and financial services are offering new revenue models for music creators, contextualised within a broader conception of market economics and value systems, the chapter further explores the ecological nature of music by reflecting on how distribution and income generation mechanisms not only support musicians' creative practice, but also inform, and to an extent shape, the nature of that practice.

Chapter 4 explores music in relation to emergent paradigms of digital creativity, exploring how music making, and the experience of listening to and consuming music, is evolving, as platforms and Internet architectures offer varying opportunities for collaborative and participative creativity. The lockdowns that were imposed worldwide as attempts to lessen the spread of the COVID-19 virus themselves catalysed a range of creative responses, not least amongst the international music community. With a growth in live-streamed music concerts during 2020–21 and the prospect of a sustained – if reduced – interest in virtual concerts as the pandemic continues to recede, it is clear that audiences have become accustomed to engaging with live music in new ways. Indeed, according to MIDiA Research, the UK-based music and media research and analysis company, virtual concerts stand to generate up to $5.2 bn annually by 2028, with an emergent generation of artists positioned to use live streaming to either supplement or replace income from traditional concert tours of small venues, using what is described as a hybrid 'URL-IRL' model. Similarly, with the launch of a number of so-called 'Creator Funds' in 2020 and 2021

(Adegbuyi, 2021), major social media platforms, including Facebook and Instagram, YouTube, TikTok and Snapchat developed new financial mechanisms to monetise and reward user-generated video content. Describing 2022 as 'the year of the creator', Mulligan et al. (2021b) indicated that music creators would engage with a range of 'post-IP solutions', referring to how traditional income streams for music, based on revenues generated by intellectual property, are increasingly being supplemented, and to an extent supplanted, by a range of alternative remuneration opportunities offered by streaming and social media platforms (Mulligan et al., 2021a).

Chapter 5 conducts an exploration of music in relation to developments in artificial intelligence (AI) in order to consider current and potential impacts on, and opportunities for, music production and creativity. By looking to current developments in AI and beyond, the chapter proceeds along two lines of enquiry: exploring the consequences for musicians as creative labourers, and developing a holistic and system-based view of music and intelligence in terms of what could be called artificial creativity.

Initially, I engage with challenging questions on the nature of human creativity in relation to music. While implementations of AI are offering a range of solutions to support, stimulate and improve music creativity and production, the quest to establish what intelligence is, is generating a number of compelling perspectives and implications. In response, the chapter engages with perspectives on authorship, ownership, even in terms of the relationship that we humans have with our own voices, in the light of artificial production and creativity. The focus then shifts, as we consider music production as creative labour in relation to debates around the future of work. The economist Daniel Susskind (2020) suggests that the history of technology's impact on the workplace has been characterised by its capacity to create new roles for human workers at the same time as it has made us redundant in established roles. By framing this process as 'task encroachment', he focuses on the way that technology replaces humans carrying out tasks within a job, rather than in job roles as a whole. Susskind proposes that the paradigm shift initiated by AlphaGo (2020) will have significant consequences across human society. In this context, the chapter examines the development of AI as a tool for catalysing creativity, engaging with the work of Holly Herndon and Mat Dryhurst on the Holly+ project as a means to explore the evolution of AI's role in creative labour processes. The project therefore provides insight into how our understanding of musical labour is evolving in response to developments in artificial intelligence, and at the same time, it offers a set of insights into fundamental questions around the nature of creative intelligence.

Many of the above themes are being explored as distinct lines of enquiry within the fields of economics, sustainability, philosophy, along with music and sound studies. However, by creating an ecomusicological point of convergence for them in *Ecologies of Creative Music Practice*, my intention in this book is to produce a new set of interlinked insights and contexts that presents music as a creative practice embedded in a set of interdependent, technological, material-discursive systems. In the context of this book, therefore, ecomusicology is a strategy. It is a way of understanding the interdependent, convergent and ecological nature of the world. By deploying such an ecological and interdependent lens and applying it to the analysis of music, my aim is to articulate the multiple ways in which music exists as part of, and in relation to, a dynamic set of environments, systems and contexts. In so doing, my hope for readers is that the particular approach that I take in *Ecologies of Creative Music Practice* enables the reader to approach music in a dynamic way.

Given the significant advances that were made in a range of thinking and creative practices in the twentieth century, which drew on both scientific and philosophical discourse, the book offers an opportunity to consider that what is emerging in the twenty-first century is a syncretic thinking that enables us to see the relations between ourselves, our creative practices such as music, and our planetary habitat, and technological, economic, cultural, philosophical and creative environments in a radically new way. In this regard, the following chapters present a deep ecomusicological exploration of cultural and commercial impacts of technological change; of making, sharing and consuming music. When we think of music in terms of technological change, we can use our knowledge of music to frame some of the broader questions of our time. How does music create new perspectives on the technological and environmental changes that we are living through? How has music been affected by these changes, how is it responding to these changes, and what changes does music catalyse? The act of 'mattering' music means to construct a fulcrum: a balancing point where we can observe the push and pull between how music both changes and is changed by the ecologies of which it is an interdependent part.

Note

1 The list comprised climate change, rate of biodiversity loss in terrestrial and marine environments, interference with the nitrogen and phosphorus cycles, stratospheric ozone depletion, ocean acidification, global freshwater use, change in land use, chemical pollution and the atmospheric aerosol load (Rockström, 2009).

Bibliography

Adams, J.L. 2009. *The Place Where You Go to Listen: In Search of an Ecology of Music.* Middletown CT: Wesleyan University Press.

Adegbuyi, F. 2021. 'Investing in influencers: 13 funds paying online creators for content'. Available at https://www.shopify.com/uk/blog/creator-fund (accessed February 2023).

Allen, A.S. 2013. 'Ecomusicology', *Grove Dictionary of American Music*, 2nd ed. Oxford: Oxford University Press.

Allen, A.S. and Dawe, K. 2016. *Current Directions in Ecomusicology: Music, Culture, Nature.* New York and London: Routledge.

AlphaGo. 2020. Available at www.youtube.com/watch?v=WXuK6gekU1Y (accessed August 2021).

Auslander, P. 2008. *Liveness: Performance in a Mediatized Culture.* 2nd edn. London: Routledge.

BBC. 2021. 'Coldplay: Band ready for backlash over eco-friendly world tour'. Available at https://www.bbc.co.uk/news/entertainment-arts-58898766 (accessed January 2022).

Benjamin, W. and Underwood, J.A. (2008) *The Work of Art in the Age of Mechanical Reproduction.* London: Penguin.

Bloom, M. 2020. 'Holly Herndon launches new podcast with Mat Dryhurst'. Available at https://pitchfork.com/news/holly-herndon-launches-new-podcast-with-mat-dryhurst/ (accessed March 2022).

Bonneuil, C. and Fressoz, J-B. 2017. *The Shock of the Anthropocene: The Earth, History and Us.* New York: Verso Books.

Borowy, I. 2013. *Defining Sustainable Development for Our Common Future: A History of the World Commission on Environment and Development* (Brundtland Commission). London and New York: Routledge.

Boyle, A. and Waterman, E. 2016. 'The ecology of musical performance: Towards a robust methodology', in Allen, A. S. and Dawe, K. (eds), *Current Directions in Ecomusicology: Music, Culture, Nature.* New York and London: Routledge.

Brennan, M. 2020. 'The infrastructure and environmental consequences of live music', in Devine, K. and Boudreault-Fournier, A. (eds), *Audible Infrastructures: Music, Sound, Media.* Oxford: Oxford University Press.

Brennan, M. and Devine, K. 2020. 'The cost of music', *Popular Music*, 39(1), pp. 43–65. doi:10.1017/S0261143019000552

Brennan, M., Scott, J. C., Connelly, A. and Lawrence, G. 2019. 'Do music festival communities address environmental sustainability and how? A Scottish case study', *Popular Music*, 38(2), pp. 252–275. doi:10.1017/S0261143019000035

Bryant, L. 2016. 'For a realist systems theory: Luhmann, the correlationist controversy and materiality', in Avanessian, A and Malik, S. (eds), *Genealogies of Speculation: Materialism and Subjectivity Since Structuralism.* London and New York: Bloomsbury.

Bryant, L., Srnicek, N. and Harman, G. 2011. 'Towards a speculative philosophy', in Bryant, L., Srnicek, N. and Harman, G. (eds), *The Speculative Turn: Continental Materialism and Realism.* Melbourne: re.press.

Cirisano, T. 2022. 'State of the live streaming nation: Reframing the format post-lockdown'. Available at https://www.midiaresearch.com/reports/state-of-the-live-streaming-nation-reframing-the-format-post-lockdown?preview=true (accessed March 2022).

Coole, D. and Frost, S. 2010. *New Materialisms: Ontology, Agency, and Politics.* Durham and London: Duke University Press.

Davies, J. 2019. 'Interdependence, or how I learned to love again on the dancefloor'. Available at https://www.factmag.com/2019/12/20/interdependence-or-how-i-learned-to-love-again-on-the-dancefloor/ (accessed March 2022).

Devine, K. 2015. 'Decomposed: A political ecology of music'. *Popular Music*, 34(3), pp. 367–389. doi:10.1017/S026114301500032X..

Devine, K. 2019. *Decomposed: The Political Ecology of Music.* Cambridge MA and London: MIT Press.

Escobar, A. 2015. *Designs for the Pluriverse: Radical Interdependence, Autonomy, and the Making of Worlds.* Durham and London: Duke University Press.

Geerts, E. 2016. 'Ethico-onto-epistem-ology', in *COST Action IS1307 New Materialism: Networking European Scholarship on 'How Matter Comes to Matter'.* Available at https://newmaterialism.eu/almanac/e/ethico-onto-epistem-ology.html (accessed September 2021).

Green, M. 2016. 'The Welsh Uplands – death or resurrection?' Available at https://www.rewildingbritain.org.uk/blog/the-welsh-uplands-death-or-resurrection#_ftn1 (accessed January 2022).

Greenberg, J.B. and Park, T.K. 1994. 'Political ecology'. *Journal of Political Ecology*, 1 (1), pp. 1–12. doi:10.2458/v1i1.21154.

Haraway, D. 2016. *Staying with the Trouble*. Durham, NC: Duke University Press.

Harman, G. 2011. 'On the undermining of objects: Grant, Bruno, and radical philosophy', in Bryant, L., Srnicek, N. and Harman, G. (eds), *The Speculative Turn: Continental Materialism and Realism*. Melbourne: re.press.

Herndon, H. 2021. 'Holly+ ? ? ?'. Available at https://holly.mirror.xyz/54ds2IiOnvthjGFkokFCoaI4EabytH9xjAYy1irHy94 (accessed March 2022).

Herzogenrath, B. 2012a 'Introduction', in Herzogenrath, B. (ed), *The Farthest Place: The Music of John Luther Adams*. Boston: Northeastern University Press.

Herzogenrath, B. 2012b 'The weather of music: Sounding nature in the twentieth and twenty-first centuries', in Herzogenrath, B. (ed), *The Farthest Place: The Music of John Luther Adams*. Boston: Northeastern University Press.

Hoffman, J. 2012. 'Q&A Bernie Krause: Soundscape explorer', *Nature*, 485, May 2012. Available at https://www.nature.com/articles/485308a (accessed October 2021).

Interdependence. 2022. Available at https://interdependence.fm (accessed March 2022).

Keychange. 2022. 'Keychange presents Pledge Action Plan and next steps towards achieving gender balance in the global music industry'. Available at https://www.keychange.eu/about-us/news-feed-articles/keychange-presents-pledge-action-plan-and-next-steps-towards-achieving-gender-balance-in-the-global-music-industry (accessed February 2023).

Krause, B. 2012. *The Great Animal Orchestra: Finding the Origins of Music in the World's Wild Places*. London: Profile Books.

Leech, O. 2021. 'Bitcoin White Paper celebrates 13th birthday'. Available at https://www.coindesk.com/tech/2021/01/21/what-is-the-bitcoin-white-paper/ (accessed March 2022).

Library of Congress. 2022. 'History of the cylinder phonograph'. Available at https://www.loc.gov/collections/edison-company-motion-pictures-and-sound-recordings/articles-and-essays/history-of-edison-sound-recordings/history-of-the-cylinder-phonograph/ (accessed March 2022).

Little, P.E. 2007. 'Political Ecology as Ethnography: A Theoretical and Methodological Guide', *Horizontes Antropológicos*, 3. Available at: http://socialsciences.scielo.org/scielo.php?script=sci_arttext&pid=S0104-71832007000100012&lng=en&tlng=en (accessed February 2022).

Live Nation. 2019. 'Live Nation sets sustainability goals for concerts and events as part of ongoing Green Nation programme'. Available at https://www.livenationentertainment.com/2019/05/live-nation-sets-sustainability-goals-for-concerts-and-live-events-as-part-of-ongoing-green-nation-program/ (accessed September 2021).

Live Nation. 2021. 'Live Nation announces Green Nation touring program, giving artists tools to reduce the environmental impact of tours'. Available at https://www.livenationentertainment.com/2021/04/live-nation-announces-green-nation-touring-program-giving-artists-tools-to-reduce-the-environmental-impact-of-tours/ (accessed September 2021).

Lousley, C. 2017. 'Global futures past: Our common future, postcolonial times, and worldly ecologies'. *Resilience: A Journal of the Environmental Humanities*, 4(2–3), 21–42. doi:10.5250/resilience.4.2-3.0021

Lovelock, J. 2019. *Novacene: The Coming Age of Hyperintelligence*. London: Penguin.

Mallin, M.A. and Cahoon, L.B. 2020. 'The hidden impacts of phosphorus pollution to streams and rivers'. *BioScience*, 70(4), pp. 315–329, doi:10.1093/biosci/biaa001

Meadows, D. 2008. *Thinking in Systems: A Primer*. White River Junction, VT: Chelsea Green Publishing.

Meillassoux, Q. 2008. *After Finitude*. London and New York: Continuum.

Mellers, W. 2001. *Singing in the Wilderness: Music and Ecology in the Twentieth Century*. Chicago, IL: University of Illinois Press.

Meta. 2021. 'Investing $1 billion in creators'. Available at https://about.fb.com/news/2021/07/investing-1-billion-dollars-in-creators/ (accessed March 2022).

Monbiot, G. 2014. *Feral: Rewilding the Land, the Sea and Human Life*. Chicago: The University of Chicago Press.

Moore, J. 2015. *Capitalism in the Web of Life*. London and New York: Verso.

Morton, T. 2010. *The Ecological Thought*. Cambridge and London: Harvard University Press.

Morton, T. 2019. *Humankind: Solidarity with Non-Human People*. London and New York: Verso.

Mulligan, M., Severin, K. and Thakrar, K. 2021a. 'Live streamed concert demand: Ready for prime time'. Available at https://www.midiaresearch.com/reports/live-streamed-concert-demand-ready-for-prime-time (accessed March 2022).

Mulligan, M., Mulligan, T., Severin, K., Kahlert, H., Das, S., Thakrar, K. and Cirisano, T. 2021b. '2022 MIDiA predictions: The year of the creator'. Available at https://www.midiaresearch.com/reports/2022-midia-predictions-the-year-of-the-creator (accessed March 2022).

Naess, A. 1973. 'The shallow and the deep, long-range ecology movement. A summary'. *Inquiry*, 16(1–4), pp. 95–100, doi:10.1080/00201747308601682.

Negarestani, R. 2018. *Intelligence and Spirit*. Falmouth and New York: Urbanomic and Sequence Press.

Pappas, V. 2021. 'Introducing the $200M TikTok creator fund'. Available at https://newsroom.tiktok.com/en-us/introducing-the-200-million-tiktok-creator-fund (accessed March 2022).

Pedelty, M. 2012. *Ecomusicology: Rock, Folk, and the Environment*. Philadelphia, PA: Temple University Press.

Pidd, H. 2019. 'Massive Attack to help map music industry's carbon footprint'. Available at https://www.theguardian.com/music/2019/nov/28/massive-attack-to-help-map-music-industrys-carbon-footprint (accessed February 2023).

Petrusich, A. 2020. 'The day the music became carbon-neutral'. Available at https://www.newyorker.com/culture/culture-desk/the-day-the-music-became-carbon-neutral (accessed February 2023).

Provenzano, C. 2018. 'Auto-tune, labor, and the pop-music voice', in Fink, R.W., Latour, M. and Wallmark, Z. (eds), *The Relentless Pursuit of Tone: Timbre in Popular Music*. New York: Oxford University Press.

Raworth, K. 2017. 'Why it's time for doughnut economics', *IPPR Progressive Review*, 24 (3), pp. 216–222. doi:10.1111/newe.12058

Richerson, P.J., Boyd, R. and Bettinger, R.L. 2001. 'Was agriculture impossible during the Pleistocene but mandatory during the Holocene? A climate change hypothesis'. *American Antiquity*, 66(3), 387–411. doi:10.2307/2694241

Robbins, P. 2020. *Political Ecology: A Critical Introduction*. Hoboken, NJ: John Wiley & Sons.

Rockström, J. 2009. 'A safe operating space for humanity'. *Nature*, 461(7263), pp. 472–475. doi:10.1038/461472a

Schafer, R.M. 1977. *The Soundscape: Our Sonic Environment and the Tuning of the World*. Rochester, VT: Destiny Books.

Schick, S. 2012. 'Strange noise, sacred places', in Herzogenrath, B. (ed.), *The Farthest Place: The Music of John Luther Adams*. Boston: Northeastern University Press.

Schwab, K. 2017. *The Fourth Industrial Revolution*. London: Portfolio Penguin.

Sharma, K. 2015. *Interdependence: Biology and Beyond*. New York: Fordham University Press.

Shimoni, D. 2012. 'songbirdsongs and Inuksuit: Creating an ecocentric music', in Herzogenrath, B. (ed.), *The Farthest Place: The Music of John Luther Adams*. Boston: Northeastern University Press.

Singer, A. 2021. 'Introducing the YouTube shorts fund'. Available at https://blog.youtube/news-and-events/introducing-youtube-shorts-fund/ (accessed March 2022).

Snap Inc. 2020. 'Snap Inc. launches Spotlight, a new entertainment platform for user generated content within Snapchat'. Available at https://investor.snap.com/news/news-details/2020/Snap-Inc.-Launches-Spotlight-a-New-Entertainment-Platform-for-User-Generated-Content-within-Snapchat/default.aspx (accessed March 2022).

Srnicek, N. 2017. *Platform Capitalism*. Cambridge, Malden: Polity.

Stockholm Resilience Centre. 2022. 'Johan Rockström'. Available at https://www.stockholmresilience.org/meet-our-team/staff/2008-01-16-rockstrom.html (accessed February 2022).

Subcommission on Quaternary Stratigraphy. 2019. 'Working group on the "Anthropocene"'. Available at http://quaternary.stratigraphy.org/working-groups/anthropocene/ (accessed February 2022).

Susskind, D. 2020. *A World Without Work: Technology, Automation and How We Should Respond*. London: Allen Lane/Penguin Random House.

Truax, B. 1984. *Acoustic Communication*. Norwood, New Jersey: Ablex Publishing.

UNESCO. 2021. 'Sustainable development'. Available at https://en.unesco.org/themes/education-sustainable-development/what-is-esd/sd (accessed March 2022).

United Nations. 2015. 'The 17 goals'. Available at https://sdgs.un.org (accessed February 2022).

United States Environmental Protection Agency. 2023. 'GHG inventory development process and guidance'. Available at https://www.epa.gov/climateleadership/ghg-inventory-development-process-and-guidance (accessed February 2023).

United States Environmental Protection Agency. 2022. 'Overview of greenhouse gases'. Available at https://www.epa.gov/climateleadership/ghg-inventory-development-process-and-guidance (accessed March 2022).

Vermaak, W. 2021. 'What is Web 3.0?' Available at https://coinmarketcap.com/alexandria/article/what-is-web-3-0 (accessed September 2021).

Villa-Lobos, H. 2015. *Villa-Lobos Piano Works: Prole do Bebê, Rudepoema and As três Marias*. Parlophone Records, Warner Music Group.

World Commission on Environment and Development. 1987. *Our Common Future*. Available at https://sustainabledevelopment.un.org/content/documents/5987our-common-future.pdf (accessed September 2021).

1

CRITICAL PERSPECTIVES ON MATTERING MUSIC

One of my core proposals in this book is that music is interdependent. It shapes what it comes into contact with, and is itself shaped by these interrelations. Indeed, at any moment, music *is* its interrelations. It is because of this interdependence that I am framing the approach to thinking about music that I take in this book as an ecological one; and why, therefore, that this book is an eco-musicological project. The function of this chapter is to set the scene; to lay the foundations of what is to come in the rest of the book. It is designed to explore and respond to a range of philosophical and critical positions that have emerged in recent years which have sought to reset how we think about the things we experience in the world around us; indeed, much of what I engage with in this chapter has been written as a fundamental challenge to decades of philosophical and critical hegemony. As Robin James wrote in *The Sonic Episteme* (2019), a certain mode of thinking has been established across a range of contemporary philosophical, political, cultural and scientific perspectives, a mode that the title of her book makes reference to, and questions. James' goal was to expose what she saw as an unrecognised, but ultimately pernicious, tendency across a range of disciplines to develop universal analytical and critical frameworks that are, mistakenly, based on what she terms a theory of 'acoustic resonance'. For James, therefore, the sonic episteme is a way of knowing that wrongly uses notions of universal harmony, and the harmonising results of statistical analyses of universal noise, to justify problematic assertions of social and cultural equality within contemporary neoliberal and biopolitical power structures.

James' work rightly highlights an invisible, or unheard, exceptionalism which lies at the heart of this sonic episteme; the idea that universal harmony and acoustic resonance are unproblematic appoaches to understanding everything that we experience. James' point, however, was that while acoustic resonance presents an alluring way to understand everything, it cannot explain everything,

DOI: 10.4324/9781003225836-2

and certain things resist the model. She suggests that neoliberal and biopolitical structures use 'statistical forecasting' (James, 2019: 27) to systematise chance events to exert power, her contention being that the analysis of statistical 'noise' has come to be the dominant paradigm for contemporary critical enquiry, which she saw as representing neoliberal, unequal and exceptionalist power structures; observing with reference to the Pet Shop Boys' song, 'Love Is A Bourgeois Construct', that 'noisy sonic distortion is the most bourgeois thing ever' (ibid.: 24). In response to such ingrained power structures, and redolent of how Deleuze and Guattari refer to the way that 'minor literature' departs from and resists the overarching narratives of majoritarian literature (Deleuze and Guattari, 1986), James' strategy was to bring to the surface a range of ways of knowing that resist the hegemonic structuring of the sonic episteme. For example, she referenced the work of Devonya Havis, whose 'study of black women's philosophical methods' (ibid.: 75) enabled her to understand how the technique 'sounding' emerged from black women's lived experience 'as a call to consider the framework and context from which [one's] actions or choices issued' (ibid.: 76). James' view of sounding as a technique and a mode of understanding that sits outside of recognised modes of academic thought clearly emphasises her commitment to acknowledging and learning from non-elite perspectives. By aggregating vernacular and non-elite ways of knowing, as expressed through a range of music and sound practices, James developed a credible and coherent set of alternatives to hegemonic structures, and it is with a similar spirit that this chapter proceeds.

In writing this book, my intention has been to create connections between music and music-related practices, perspectives on technology, and vectors in contemporary philosophical and ecological thought. However, as I said in the Introduction, this involves bringing together a number of sometimes competing and conflicting perspectives. The work of this chapter is not to create a framework that resolves these conflicts; instead, I am interested in what we can learn about music and its relation to a wider ecosystem of technological, commercial and philosophical trajectories, and at the same time explore the consequences of music's entanglement with these wider forces. So saying, the prime function of this chapter is to engage with a set of critical paradigms that centre on deep ecology, new materialism, speculative realism and non-philosophy in order to establish and account for the book's overarching ecomusicological approach. We begin with Arne Naess' formulation of Deep Ecology, a concept and an approach to thinking which provides the fundamental framework for my use of the interrelated principles of ecology and interdependence throughout this book. We then encounter an array of philosophical perspectives that, in their distinct ways, have been developed to articulate the fundamentally material and interrelated substance of the world, and which suggest a number of routes into thinking about music's interdependent relationship with technology, commerce and practices of making, listening and sharing. As a result, and following Naess' articulation of the connection between ontology and ethics (Naess, 2008: 77), we are more able to consider the implications for creativity and music-related practices in terms of these ontological models of interrelation.

Deep Ecology

In the short essay 'The World of Concrete Contents', the ecologist and philosopher Arne Naess sets out an ontological model that underpins his conception of deep ecology. One of the things that we can take from Naess is that things really are a part of each other. Whereas post-structural theory might have boiled the nature of experience down to a solipsistic gruel – which Quentin Meillassoux reformulates as 'correlationism' – a challenge that Naess' deep ecology lays down is to think in terms of objective reality, but to understand that an objectively existing world simultaneously contains humans, and is contained by humans. In other words, for Naess, humans really are part of the world, and the world really is a part of humans. Whereas Meillassoux's use of the concept of correlationism draws on a post-Kantian trajectory that essentially frames experience in terms of the capacity to have experience, Naess works towards the view that human experience is conditioned by a non-subjective relationship between the individual and the material world. In this sense, while the subjective human capacity to experience the world shapes our understanding of the world, this capacity for experience is fundamentally shaped by an objectively existing world, and is not simply a component part of a self-created and subjectively limited perspective. It is this reciprocity between an objectively existing world and human knowledge and experience that has been extensively explored by a range of critical and philosophical thinkers in the twenty-first century, and which can be seen as a fundamental commonality across speculative realist and new materialist trajectories.

Naess expresses this idea in the following way:

> Essential to ecological thinking, and to thinking in quantum physics, is the insistence that things cannot be separated from what surrounds them with smaller or greater arbitrariness. Thing A cannot be thought of in and of itself, because of internal relation to thing B. But neither is thing B separable, except superficially, from thing C, and so on.
>
> *(ibid.: 72)*

As we shall see, such a view has formed the basis of a great deal of examination and debate, particularly during the early decades of the twenty-first century.

Naess' own conclusions as to how interrelatedness and interdependence can be reconciled in terms of theories of an objectively existing world, as against the all-encompassing nature of subjective experience, are expressed in his articulation of how a bowl of water can be experienced as having different temperatures by a person dipping their hands in separately. The first hand, after having been kept cold, experiences the water as warm, while the second, having been kept at room temperature, experiences the water as cold. Naess' strategy is to propose that things have no absolute, primary qualities in and for themselves. In terms of the perceived warmth of the water in his example, he suggests that the labelling of the water as being warm 'refers to water in relation to a complex set or constellation of relata, of which the most obvious are the hand, the water, the medium, and the subject's uttering "Warm!"' (ibid.: 72).

This is an important point as regards Naess' deep ecological ontology. For him, these 'relata' are not separate, individual objects or entities. Although they have individual names, they are internally related to each other as part of what he refers to as an 'indivisible structure [or a] constellation of factors' (ibid.: 72). The consequence of this is that, in terms of the example of perceived water temperature, for Naess, there is no separately identifiable water or organism that is related to the other; the relationship is between these relata – or factors – and the whole of the constellation. Thus, the act of feeling something does denominate the experience of reality; humans are in reality anyway, touching water, or feeling either its warmth or coldness, is not a marker of reality. As he says, 'it is misleading to call it real only as felt by a subject' (ibid.: 73). As we shall see, Naess proposes that it is the concrete content of the world as an indivisible structure, along with the abstract structures that subdivide the constellation into named units by labelling things as 'water', 'warm', 'cold', 'hand' and 'bowl' that comprise reality.

Naess' use of the concept of a 'constellation of factors' provides us with a key to understanding how an objective/concrete reality exists that we are part of, and reciprocally, is a part of us. His idea that the relation between the relata – in other words the entanglement between all things – is internal, means that we don't experience the relation as something that we have a feeling or an opinion about; the entanglement is happening at a fundamental, structuring level. The entanglement with relata constructs our capacity to have experience, but it also constructs the capacity of relata to be experienced. Naess' point here is that the constructing-function of entanglement is precisely that; it is not a matter of subjective experience, it is a concretely existing function of the relation, not of the experiencing subject.

Naess arrives at the statement:

> I maintain that the framework of gestalt ontology is adequate, but scarcely the only adequate one, in any attempt to give the principles of the deep ecology movement a philosophical foundation. The world of concrete contents has gestalt character, not atomic character.
>
> *(ibid.: 80)*

Naess uses two primary concepts here to lay out his deep ecological ontology, and its relation to ethics: concrete content and abstract structures. By concrete content, he means the world that we experience, and which we are part of. He refers to the totality of the world-as-concrete-content as the gestalt structure. In regard to the example of a bowl of water that feels both warm and cold, then the bowl, the water, the opposing sensations of cold and warmth – all of these components are part of this gestalt structure, which is to say that they are all equally things that are real occurrences in the world. He is careful to clarify that the gestalt is a self-contained whole, and that we do not have subjective experience of the gestalt as such, rather that our experience – as concrete content – occurs within, and structures, the gestalt. Naess' second term, abstract structures, refers to how we interact with, and understand, the world. Again, using the example of the bowl of water, the way that we are able to understand that it is the same bowl of water that produces two different sensations is a quality of abstract structuring. Here, Naess is drawing on the philosophical theory known as nominalism, which proposes that the world is made up of individual things, rather than general properties or types. In this context, the word 'same' is simply a label that is being used to denote the self-sameness of the water, and is part of an abstract structure. Thus, the bowl of water, along with the warmth and the coldness of the water are all concrete contents of a gestalt structure, whereas the sameness of the water is part of an abstract structure laid on top of the gestalt structure that enables humans to interpret and organise the world. To this end, Naess suggests that 'the world has structures, but does not reveal them. We make conceptual constructs to cope with them, but they are all human-made. Gravity does not pull planets!' (ibid.: 78). With this, Naess signals that although the force of gravity is something that exists as a concrete content of the world, and operates as part of what he calls the gestalt structure's 'internal structural relations' (ibid.), the term 'gravity' itself is part of the abstract structure of science, as, too, are the concepts 'ecosystem' and 'deep ecology'. While abstract structures are seen as a necessary means of organising the world into something humans are able to navigate and successfully inhabit, he concludes his analysis by stating that 'abstract structures are of the world, not in the world' (ibid.: 79), which is to say that they do not ever become part of the concrete contents of the world, they are merely overlaid, almost as a means of orienting ourselves.

To return to Naess' concept of deep ecology, then, although he frames it as part of an abstract structure designed to orient our relationship with the world, it is in a wider sense fundamentally an ontological model that comes with an ethical imperative. Naess draws on an example from the natural world – trees – in order to communicate the idea of ontology as ethics. He positions an environmental activist and a developer as two poles in a debate over the value of trees, and suggests that their differences as regards their views on trees are essentially differences of opinion about what constitutes the real world. For Naess, the activist may well view a forest as a gestalt whole that will be irremediably damaged if large parts of it are cut down, whereas the developer sees that cutting down a certain number of trees through the middle of the forest in order to construct a road is a reasonable intervention. The question of seeing the world as a gestalt whole, wherein everything is an interrelated part of its concrete contents, is therefore a question of ethics; that disrupting gestalt structures have significant consequences.

Similarly, one of the ethical questions that Naess raises, as part of his discussion of the concreteness of the world, is a question of value. If our ontological model separates us from the world that we inhabit (by which I mean the world as gestalt structure, not simply the 'natural world' of flora and fauna, which is itself an abstract structure), then 'there is no good reason why we should not look upon such a bleak nature as just a resource' (ibid.: 74). From a subjective – correlationist – perspective, it would seem that the world has no inherent qualities, that value is something that we subjectively confer on the world as something separate from ourselves. Here, a world without inherent qualities is a world without inherent value: value is in the eye of the beholder.

In 1984, Naess set out what he referred to as a 'common platform' of deep ecology, so as to distinguish it from the various philosophies and religions that informed it, preferring the word 'platform' to 'principles' so as to avoid the points themselves becoming too rigid or a set of commandments to be dogmatically adhered to (ibid.: 105). The eight points of the platform comprised a set of values and proposals for implementing a deep ecology viewpoint into human concerns and activities. The first point, which to an extent amounts to an overarching principle, was that 'the well-being and flourishing of human and nonhuman life on Earth have value in themselves .... These values are independent of the usefulness of the nonhuman world for human purposes' (ibid.: 111). We saw this position emerge in his discourse on concrete content, where he laid out his philosophical view that human experience was not the criterion of existence or value for nonhuman contents of the world, or indeed the world itself. Although Naess deliberately avoided using the word 'principles', the platform does, however, also set out a series of ethical principles. These include the idea that the richness and diversity of the world is a value in itself, that working to achieve a better quality of life is not the same as simply striving for a higher standard of living, and that humans do not have the right to reduce the diversity of non-human life on earth 'except to satisfy vital needs' (ibid.). Naess also suggests that although developing human cultures is compatible with reducing the global human population, enabling non-human life to flourish can only happen if the number of humans on the planet decreases. The sixth point is a general policy recommendation, which states that in order to achieve a higher quality of life for humans and non-humans, then changes need to be brought about in the way that economic, technological and ideological structures are managed and coordinated.

It is worth noting that Naess was clearly aware of his critics' views, quoting the concern that the deep ecology movement was too focused on 'the biosphere as an organic whole' rather than on human and non-human life (ibid.: 107). Naess' response to this perspective was that the notion of an organic whole was a metaphysical concept, rather than a concrete reality, and that the deep ecology eight-point platform was an interface between philosophical and religious principles relating to such metaphysical conceptions and the concrete content of the world in terms of policies, factual hypotheses, rules, decisions and actions (ibid.: 107–9). In this context, deep ecology has also had its supporters, including the physicist Fritjof Capra, who has written extensively about systems thinking (2014), and the importance of developing a scientifically informed systems approach to understanding and ecologically balancing current and future existence on earth (1996, 2014). Capra draws on Naess' work directly to connect his own use of the word 'ecological' with the deep ecology movement, and differentiates shallow ecology – which positions 'humans as above or outside of nature' – from deep ecology – which sees the world 'not as a collection of isolated objects, but as a network of phenomena that are fundamentally interconnected and interdependent' (Capra, 1996: 7). Indeed, the title of Capra's 1996 book *The Web of Life* springs directly from the vision of interconnectivity and interdependence that Naess sets out in his own work. In addition, Capra's own view about Naess' deep ecology ethics is worth considering in the light of the criticisms levelled at Naess over his perceived sidelining of human needs for the sake of the organic whole. For Capra, the shift that deep ecology brings in values is a decentring of human values and a move towards 'ecocentric' (earth-centred) values. Capra's point here is that, rather than removing, or even displacing human values, the work of deep ecology is to foreground human involvement in a complex 'web of life', with the view that 'all living beings are members of ecological communities bound together in a network of interdependencies' (ibid.).

Naess' deep ecology is therefore a framework for a materialist ontology, at the same time as laying the foundations for an ethical orientation towards the world. It is a framework based on a philosophical view that encompasses – in his words – a metaphysical conception of the real world that produces a set of ethical imperatives. While we do not have direct experience of the world in Naess' metaphysical sense of it, we are nonetheless completely integrated into, and part of, its interdependent structure; our decisions and actions have a real impact on the world, and our existence is co-extensive with a range of human, non-human and inanimate companions.

Speculative Realism

The Speculative Realism Workshop, held at Goldsmiths College in 2007, brought together four thinkers – Ray Brassier, Iain Hamilton Grant, Graham Harman and Quentin Meillassoux – whose philosophical interests had sufficient commonality to create a meaningful discourse between them. As with any movement, or indeed any musical genre, the moment it is given a name its key adherents seem to want to have nothing to do with it. For Brassier,

> The term 'speculative realism' was only ever a useful umbrella term, chosen precisely because it was vague enough to encompass a variety of fundamentally heterogeneous philosophical research programmes [...] There is

> no 'speculative realist' doctrine common to the four of us: the only thing that unites us is antipathy to what Quentin Meillassoux calls 'correlationism' – the doctrine, especially prevalent among 'Continental' philosophers that humans and world cannot be conceived in isolation from one other – a 'correlationist' is any philosopher who insists that the human-world correlate is philosophy's sole legitimate concern.
>
> *(Brassier, 2009)*

Taking a wider view, Armen Avanessian and Suhail Malik, writing in *Genealogies of Speculation* (2016), see that the common cause that united the so-called 'speculative realists' was a shared desire to move beyond the strictures of post-structuralism[1] and analytic philosophy.[2] Both were key moments in twentieth-century philosophical and critical development, but reflected two very distinct schools of thought. The concept of correlationism that Brassier referred to was a legacy of the work of Enlightenment philosopher Immanuel Kant. In their own way, each of the workshop delegates saw that Kant's ideas have continued to inform much – if not all – of the currents in mainstream Western philosophy up to the present day. Although it was only Meillassoux who specifically referred to the philosophical tendency to think of the world as the product of human subjective experience as 'correlationism', the delegates' shared ambition was undoubtedly to argue that it must be the case that a real world exists beyond a human capacity to experience or conceptualise it. Since 2007, speculative realism has come to represent a certain orientation in thought within continental philosophy, leading a variety of thinkers to re-examine traditional notions of post-Kantian ontology and epistemology which in itself has been referred to as the 'speculative turn' (Bryant, Srnicek and Harman, 2011). Fundamentally, the notion that human thought is able to conceive of an absolute, concretely existing 'real' world, which is not just the product of human thought was the premise that gave the workshop, and the subsequent movement, its name.

Meillassoux's book, *After Finitude: An Essay on the Necessity of Contingency*, explored this premise extensively, and one of his signal contributions that catalysed and defined the speculative turn was his proposal that an absolute reality – which he referred to as a 'great outdoors' – exists beyond human thought; a real that we can conceive of, but a real that is absolutely not the product of human thought (Meillassoux, 2008a: 7). According to Alexander Galloway, central to Meillassoux's project is the notion of 'existence without givenness' (Galloway, 2010: 8), which is to say that, whether or not humans are there to experience it, an absolute reality exists (in other words, reality is not just 'given' or presented through human experience). The concept of the absolute is another key reference point that runs throughout speculative realist discourse. In *Cartographies of the Absolute*, Alberto Toscano and Jeff Kinkle define the absolute as a philosophical term that is used to refer to an uncategorisable register of reality. For Toscano and Kinkle, the absolute 'defies representation' – certainly in terms of human capacity to perceive, comprehend or conceptualise it. What is more, due to its roots in theological thinking, the absolute is also understood to be that which is 'infinite and unencompassed' (Toscano and Kinkle, 2014: 23). In this sense the 'finitude' in the title of Meillassoux's book is a direct reference to this uncategorisable infinite, and the phrase 'after finitude' gestures towards an absolute that cannot be encompassed by the capacity of human thought to conceive of it.

Although in subsequent years the debates have moved on, the key tenets of speculative realism still inhabit the work of various contemporary thinkers, including Reza Negarestani (2014, 2018) and Timothy Morton (2013, 2018; Morton and Boyer, 2021), whose work continues to investigate the complex relationship between human thought and the world of lived experience. There is much that can be said about, and in the name of, speculative realism, however, in the context of *Ecologies of Creative Music Practice*, my interest is to engage with Meillassoux's concept of correlationism. By exploring his use of the term, the philosophical obstacles that he was trying to overcome, and the solutions he designed to overcome these obstacles, we can observe an essential aspect of the speculative realism 'project', and look at how we can connect it with wider issues of music and the ecomusicological focus of this book. Meillassoux informs us that, by '"correlation" we mean the idea according to which we only ever have access to the correlation between thinking and being, and never to either term considered apart from the other' (Meillassoux, 2008a: 5). His target is what he calls 'post-critical thought' (ibid.: 4), a philosophical approach to thinking which he sees as being based on the idea that, if anything can be said to exist, then it must exist solely in terms of that which perceives its existence. For Meillassoux, post-critical thinking requires us to accept that 'it is naive to think we are able to think something' that exists in its own right (ibid.). In the case of human thought, therefore, something can exist only in terms of our capacity to perceive, think about or imagine it.

Another consequence of thinking about human experience solely in terms of a correlation between objects of thought and the human capacity to think them is that we quickly reach an impasse, where not only must everything that exists be thinkable, but, more problematically, only that which is thinkable can exist. In opposition to this, Meillassoux's intention is to construct a substantial and logical proof for an autonomous absolute that sits outside the confines of what he sees as the 'thesis of the essential inseparability of the act of thinking from its content' (ibid.: 36), where 'all we ever engage with is what is given-to-thought, never an entity subsisting by itself' (ibid.). *After Finitude*'s project is therefore to demonstrate that an absolute – one that is external to human thought – must exist, which by definition would prove that a correlation between thought and the world is not the limit of human thought.

Meillassoux's own response to this paradox of looking for something that we cannot recognise is to undermine what he sees as the finitude of the correlation in two ways, and he conceives of two strains of the correlationist perspective: a 'weak' and a 'strong' correlationism in order to do this. Weak correlationism is for Meillassoux the Kantian formulation of the phenomenal and noumenal realms, where Kant reasoned that there is a world of experience (the phenomenon) as well as a world of the in-itself (the noumenon), but that it is logically impossible for humans to have any access to this noumenon. Strong correlationism is the process of 'absolutising the correlation in itself' (ibid.: 37), which is to say that the correlation between thought and, as we have seen, what is thought about, becomes the only thing that can be thought (in other words, this variant will not allow for even the consideration of a noumenon in the way that Meillassoux's reading of Kantian thought permits). Fundamentally, Meillassoux set up strong correlationism as a perspective that de-legitimises any attempt to say anything about anything that exists beyond the reach of human thought; everything beyond thought is literally inconceivable, and simply cannot be discussed. As such, the correlationist model would compel us to accept that what is given in experience is fundamentally given for us, and that anything that we imagine to be beyond what is given is precisely that: simply a product of our imagination. As far as Meillassoux is concerned, correlationism means that any conception of something beyond what is given in experience is actually only the result of something that is already given in thought, and therefore trapped within a feedback loop of finite thought.

If the focus of *After Finitude* was on defining and then destabilising the concept of correlationism, then chief among Meillassoux's philosophical tools are ideas relating to contingency and facticity. He uses these two terms to put forward his key claim about a post-critical ontology. First, contingency describes how the physical laws that govern absolute reality are indifferent to any events and phenomena that may occur. In this, Meillassoux proposes that contingency is the only quality of the real world that is necessary:

> [We] know two things [...]: first, that contingency is necessary, and hence eternal; second, that contingency alone is necessary [...] Everything is possible, anything can happen – except something that is necessary, because it is the contingency of the entity that is necessary, not the entity.
>
> *(ibid.: 65)*

On this platform, Meillassoux then places the concept of facticity, which essentially suggests how we are unable to make any account for why anything in the world of cognitive experience exists. For Ray Brassier, 'facticity [...] pertains to the principles of knowledge themselves, concerning which it makes no sense to say either that they are necessary or that they are contingent, since we have no other principles to compare them to' (Brassier, 2007: 66). Essentially, facticity describes the situation we find ourselves in as inhabitants of a world we have no way of understanding, and no way of accessing the principles or laws that it rests upon. Galloway expands on this idea, saying that 'if contingency says that the physical laws remain indifferent to events, facticity says we can only describe these kinds of laws, we can never found them, meaning we can never think of them as absolute ontological axioms' (Galloway, 2010: 8). In this, Meillassoux's and Galloway's views echo Naess' formulation of the concrete content of the world, which is inaccessible to humans, and the abstract structures that we generate in order to navigate the concrete world. Again, the idea that we are very much in, and part of, the world, but we are not able to understand how it operates at a fundamental level; what Meillassoux would refer to as the absolute.

However, Meillassoux does move us on from Naess' position which separates concrete content from abstract structures. Facticity is a concept which 'forces us to grasp the "possibility" of that which is wholly other to the world, but which resides in the midst of the world as such' (Meillassoux, 2008a: 40). In this sense, our knowledge about the limits of our knowledge, and our inability to understand anything about the grounding of the world in which our knowledge occurs, can only occur within the world in which our knowledge exists. Although convoluted and tautological, it is important to establish Meillassoux's position here, because it will have a significant impact on how we think about creative practices of music from this point onward. In this way, facticity becomes a means to make visible and interrogate the problematic nature of thinking about the limits of thought itself, and asks us to consider how it is that we can use thought to think about something that exists beyond the confines of thought.

Meillassoux uses facticity to identify and emphasise the inherent self-contradiction of the correlationist position, by showing that the logic that supports the correlation must by definition cancel itself out. He therefore presents us with two options. On one hand, he tells us that if we accept the governing principles of facticity – if we 'absolutise' facticity – we would need to accept that our experience (and knowledge) of the world is only ever of a world that is presented to us (in other words that it is not 'reality', simply a 'presentation'), but at the same time, we can have no knowledge or understanding of how the world comes to be presented to us. As a result, this would mean that the correlation itself is 'de-absolutised'; it simply becomes a concept 'for us', part of the world that is presented to us, and an absolute truth or reality in itself. On the other hand, Meillassoux tells us, if we insist on the truth of the correlation (if we wish to absolutise the correlation), then we must simultaneously de-absolutise facticity. By asserting that 'facticity is only true for-me', Meillassoux argues that the correlation itself becomes 'nothing more than the correlate of my thought' (Meillassoux, 2008a: 59), completely shattering the fundamental precept of the correlation. His point is that, if we choose to absolutise the correlation, any statement that is made from within the correlation must itself be limited by the confines of the correlation. In other words, the correlation can only be 'true for me'. Ultimately, the correlationist's problem is that if the correlation is 'true', then it cannot be true, since any claims about the truth of a correlationist perspective would always have to come from a correlationist perspective. On the other hand, facticity, as a statement about our inability to know what underpins knowledge, is a claim that it is 'true' that there are certain things that we cannot know, but that we can be sure that we cannot know them. This in itself is a form of knowledge about the unknown, and is the subtle but critical difference between Meillassoux's and Naess' perspectives. Where Naess saw that we are in a world that we cannot comprehend, only name, Meillassoux's work is a pathway towards knowledge of the unknowable. Following Meillassoux's logic, it must be possible, from with the confines of the world that is presented to us, to have knowledge of something that is beyond those confines, even if, as Brassier says, our knowledge is that we can know nothing about what is beyond those confines (Brassier, 2007: 66). Even knowledge about a lack of knowledge of what is beyond the presented world, and an awareness that the world that we experience is simply a presentation or a correlation is in itself tangible. In addition, it is essential that we create a distinction in our thinking about the 'principles of knowledge themselves': we may have no knowledge about what these principles are or how they work, we can however confidently say that we know that we do not know these things. Although convoluted, Meillassoux's logic suggests that it must be conceivable to think about an absolute that is beyond thought but at the same time remains accessible to thought. In other words, we must be able to 'think' the 'unthinkable'. By absolutising facticity, and by demonstrating the necessary de-absolutisation of the correlation, Meillassoux is able to assert that what we might take to be the truth of our experience is in fact merely an aspect of a wider field of knowledge, which includes knowledge about the experiences we are having, alongside a knowledge that recognises our inability to know anything about the underlying principles of knowledge.

While Galloway reflects that Meillassoux's work communicates 'a kind of strong nihilism, evoking a world of chaos and radical contingency' (Galloway, 2010: 9), there is much we can take from the latter's approach that enables us to build on Naess' work, or at least, consider deep ecology from a speculative realist perspective. The interdependence that Naess finds compelling in his creation of a flat plane of existence in which humans are equally as embedded as non-humans, while different, is nonetheless comparable to Meillassoux's articulation of facticity: we are all still in a world. Similarly, Naess' twinned notions of concrete content and abstract structures bear comparison to the way that correlation is seen to operate, that we are only able to access what is given in experience. Meillassoux's contribution is to simply say that because of the factical nature of the way that human experience works – which can be paralleled with abstract structures – it must be possible for us to be able to establish some knowledge of the absolute contingency that underpins our world of given experience.

New Materialism: Agential Realism

To an extent, the ontological and epistemological theories that underpin new materialism are moving in the opposite direction to speculative realism. Whereas the likes of Quentin Meillassoux, as we have seen, focused on the necessary separation between the observer and the observed, new materialist thinking often starts from a position that, not unlike deep ecology, sees the fundamental interconnectedness and interrelatedness of thought, discourse and the material nature of things. Something that new materialism does share with speculative realism, however, is that it too is by no means a homogeneous philosophical project. In *New Materialisms: Ontology, Agency, and Politics* (2010), the collection that drew certain trajectories in thought together and provided them with a collective identity, Diana Coole and Samantha Frost acknowledge the heterogeneity of the then-emergent wave of materialist-inflected thought. Not only did the 'new materialisms' (Coole and Frost, 2010: 4) that their collection brought together engage with a diverse range of topics, the various authors drew on an equally broad range of thinkers, including 'the great materialist philosophies of the nineteenth century [...] of Marx, Nietzsche and Freud' (ibid.: 5), along with Einstein, Foucault, Bourdieu, Lefebvre, De Certeau, du Beauvoir, Merleau-Ponty, Althusser, Hobbes, Deleuze, Bergson and Spinoza. In addition, Coole and Frost count Descartes, Newton and Euclid as amongst those who – despite the new materialists' generally shared rejection of their scientific, mathematic and philosophical positions – nevertheless made a significant contribution to the conceptual environment in which these contemporary thinkers had been working.

Although their work was not presented in Coole and Frost's collection, the feminist theorist Karen Barad, whose background is in particle physics and quantum field theory, has come to be identified with new materialist thought, by theorists including Andreas Malm (2018), Paul Rekret (2018) and Robin James (2019). Barad's key work, *Meeting the Universe Halfway* (2007), used the precepts of quantum physics to put forward a materialist argument for grounding human thought and discourse in the movement of electrons. Two key terms feature in their work: entanglement, which describes how particles become enmeshed with each other such that they are no longer individually identifiable, and diffraction, the process by which colliding waves – for example, water, sound or light waves – overlap with each other to create resultant waves in increasingly complex patterns. Another term drawn from quantum physics – 'superposition' – refers equally to entanglement and diffraction, and denotes a state where it is impossible to identify either particle- or wave-like behaviour (Barad, 2007: 74 and 270).

Throughout *Meeting the Universe Halfway*, Barad's formulations drew on the work of two physicists, Niels Bohr and Erwin Schrödinger, each of whom developed now-famous theories about the central role that observation plays in the outcome of a given event (ibid.: 276). Barad references Bohr's hypothetical 'two slit' experiment, which was intended to show that, depending on whether a stream of electrons that have been fired at a screen are observed or not, they would land in a wave-like diffraction pattern, or as individual particles in a mixture pattern (ibid.: 307). Similarly, Barad acknowledges the role that Schrödinger played in refining quantum theory, citing his view that Bohr's ideas about the importance of observation on the outcome of a given situation could not hold at the level of observable experience. For Schrödinger, the thought that a hypothetical cat shut in a sealed box could be simultaneously alive and dead, in the same way that Bohr saw that entangled particles existed in superposition as a wave until our observation collapses them onto a screen as mixed particles, was 'quite ridiculous' (ibid.: 276–7).

For Barad, however, the collapse never happens. Waves do not become particles upon observation, since the world is simply a never-ending series of superpositions – diffractions and entanglements – that exist as constantly evolving topologies (ibid.: 345). Barad's view is that there is no difference between the 'quantum' world of atomic and subatomic superposition, and the world of 'classical physics' that we are able to observe at the level of human experience; it is simply that atomic and subatomic behaviours are too small for us to perceive. Therefore, given that the difference between quantum behaviour and human experience is merely a difference in scale, rather than in type, everything that exists is in a constant state of entanglement (ibid.: 279). Similarly, the theoretical physicist Carlo Rovelli's multiple quantum state theory allows for nested superpositions, in that a reality for one observer, which cannot be observed by an observer in a different vantage point, can both be real and in superposition at the same time. Rovelli reformulates Schrödinger's hypothetical cat experiment to illustrate this idea, updating the experiment slightly for a contemporary readership. The cat is now a human, sealed in a room containing a chemical preparation that will send them to sleep if the behaviour of an electron triggers the release of sleeping gas. In this new version, the human is also aware of whether they are awake or asleep. For observers outside of the sealed room, they will not know what is happening inside until they open the door and see whether the experimentee is awake or asleep. For Rovelli, this demonstrates that reality or facts can have two meanings simultaneously, depending on the position – or context – of the observer. As with Barad's assertion that quantum behaviour is always happening – it's just that we humans are not equipped with the necessary apparatus to observe the movements of electrons in the world around us – Rovelli also affirms that we are never separated from a world of quantum activity. For Rovelli, we cannot see quantum phenomena because of interference. This is to say that the reason we are unable see if Schrödinger's cat is alive or dead because of interference, in other words, we cannot see through the walls of the box that the cat is in (Rovelli, 2021: 95–6).

Although Barad is clearly aware that Bohr is not without his critics in the wider field of theoretical physics, and acknowledges certain 'ambiguities' in his work with regard to the correspondence of 'theory and reality' (Barad, 2007: 122–3), two other aspects of Bohr's work are central to Barad's ontological and epistemological modelling. Indeterminacy, the fundamental impossibility of accurately measuring the position of a particle, is at the heart of their quantum ontology. Referencing Bohr, Barad relates that it is impossible to measure a particle's position without measuring its momentum, but since the process of measuring its momentum would interfere with the particle's position – and measuring the particle's position would require halting its momentum – these two values cannot be known at the same time (ibid.: 118). Leading on from this, Barad describes complementarity as the impossibility of making any kind of measurement of atomic objects, since any measuring device will always interact, and therefore interfere, with the object that it is measuring (ibid.: 308). At stake here is the idea that, because of complementarity, it is completely impossible to make any objective measurement of, or intervention with, the world. This is Barad's epistemological point: we cannot avoid being involved in the world, no matter how hard we try. Any act of measurement will always impact on the object that is being measured. What is more, even if we were able to overcome our inability to avoid influencing the things we are measuring, indeterminacy tells us that we can only ever perceive the world from a certain perspective at the cost of invalidating other perspectives; we can never achieve a holistic view of the world as an objective, complete whole. Given Barad's views that quantum physics pertains across all scales of the world, from subatomic particles through to human experience, the problem nature of measurement must also be constant.

Having established the importance of indeterminacy and complementarity, Barad proceeds to build on this foundation, suggesting that, since it is impossible to measure anything without creating contradictions, and because there is never a definite point where indeterminacy does not exist, meaning cannot exist in and of itself. As such, meaning is always something that comes into effect once something is being measured, observed or experienced. This would suggest that, whether we are discussing a piece of music, a poem, a Nike running shoe, or the Burj Khalifa, anything that humans make – exactly like particles – does not have an objective or inherent meaning. Everything is indeterminate, and everything is impacted – or created – by the process of being experienced. Indeed, taking Barad's point further, it is also important to recognise that every observer, everyone having an experience, is themselves is entangled with the world. For someone listening to a piece of music, just as there is no absolute sonic object that they are listening to, neither is there an absolute vantage point from which they are listening. Listening itself is an entangled process: 'we are part of that nature that we seek to understand' (ibid.: 26).

As I have written elsewhere, Barad makes frequent use throughout *Meeting the Universe Halfway* of the phrase 'marks on bodies' (Barad, 2007: 232) to convey a sense of what happens when things, which can also mean things as aggregates of forces – for example people, ideas, animals, rocks, music – become entangled with, and diffract each other (Lovett, 2021). In simple terms, a 'mark' on a body – body as thing, body as force, body as discourse – is shorthand for spacetimemattering: a mark is a spacetimemattering. This relates to one of Barad's core narratives: the absolute non-neutrality of any measuring process. Whereas Meillassoux simply dismisses the correlationalist view of the observer–observed relationship, which cancels out any reality beyond the reality of the observer's capacity to observe things, Barad leads us to a more nuanced position. Measurement is a process of becoming entangled with the object of measurement, and entanglement leaves its mark on both the observer and the observed. The marks left on each body are real, and so are the interacting bodies. Everything about the process of measurement-entanglement is real. The key difference between Barad's and Meillassoux's perspectives is that Barad's agential realism is a proposal that allows for the tangible production of the real in the world. Another important aspect of this formulation is that the process of measuring – in other words, observing, listening, experiencing, and so on – not only produces meaning, but it is part of the meaning that is produced. Barad's collapsing of measurer and measured, of observer and observed, leads us by another route to their formulation of agential realism, which we can also think about in the context of creativity:

> Since there is no inherent distinction between object and instrument, the property measured cannot meaningfully be attributed to either an abstract object or an abstract measuring instrument. That is, the measured value is neither attributable to an observation-independent object, nor is it a property created by the act of measurement. [Therefore] measured properties refer to phenomena, remembering that the crucial identifying feature of phenomena is that they include 'all relevant features of the experimental arrangement'.
>
> *(ibid.: 120)*

This suggests that it is the measurement process, and the production of boundaries that constitutes the real; the fact that an experience is not observer-independent does not make it any less real. Indeed, it is only through the entangled process of measurement that the real is produced, there is no other way. Barad is simply describing – as they see it – the process of how everything in the universe is continually interacting/intra-acting with everything else.

Barad uses the term spacetimemattering to convey this fully entangled process that involves the constant production of time, space, matter and meaning. It is for this reason that we can describe Barad's ontological model as topological. Their analysis enables them to frame human experience as a process of observation, akin to the observation process in the two-slit experiment, in a dynamic relationship with a broader consideration of the nature of time. This suggests that, at a fundamental level, things are not simply made against a backdrop of constant time, but that things and time are created as part of an ever-changing topology of what Barad calls 'spacetimematter' (ibid.: 177). As such, it becomes clear that Barad's world is not a world of 'things', it is a world of movements, collisions and interactions, or what they refer to as 'intra-action', a concept that describes the deeply interdependent and interlaced nature of everything that is in the world – including humans and music. There is an obvious resonance with deep ecology here, but Barad's quantum ontology does not simply refer to a world of tangible substance, but to discursive practices as well. This is one of the most significant advances in thinking that Barad proposes in their work, the sense that, not only are the worlds of quantum behaviour and human experience the same kind of thing, but so too are the worlds of matter and discourse: each is produced by the same fundamental processes, and each is therefore able to engender appreciable change in the other.

Although Barad's work has developed along a notably different trajectory from that of Meillassoux and the speculative realists – indeed a key reference for *Meeting the Universe Halfway* is the post-structuralist thinker Michel Foucault – Barad is clearly aware of the dangers of correlationist-like solipsism. They point out that although 'there is no absolute condition of exteriority to secure objectivity [...] this doesn't mean that objectivity is lost. Rather, objectivity is a matter of exteriority-within-the-phenomenon' (ibid.: 345), an idea not dissimilar to Meillassoux's articulation of facticity, and our capacity to think the great outdoors, even though we cannot conceptualise it. On Barad's reading, an objective exteriority is very much part of the world of experience, and therefore, we are part of that exteriority, while at the same time it is an exteriority that is part of us and the things that we make. Again, this is a position that pushes beyond Naess, and Barad would see no distinction between concrete content and abstract structures; each would simply be a different expression of spacetimemattering, equally accessible to, and formative of, human experience.

Barad further describes how this interior-exteriority, objectivity-within-subjective-experience operates: 'Objectivity is a matter of accountability to marks on bodies. Objectivity is not based on an inherent ontological separability, a relation of absolute exteriority, but on an intra-actively enacted agential separability, a relation of exteriority within phenomena' (ibid.: 340).

Since exteriority is not an already-existing state, Barad uses the idea of intra-activity to convey how exteriority is produced within the context of entanglement/diffraction. Where interaction suggests that two or more self-contained bodies are engaging with each other, intra-action describes a situation where two or more bodies are produced as a result of them overlapping. Intra-action is the process of interdependent entities becoming entangled, simultaneously producing and being produced by each other (which would suggest that arriving at any kind of objectively-existing point, or moment, of exteriority is impossible), and where everything that is happening could not exist anywhere else in space and time (because even space and time are co-produced through the process of intra-action). The significance to Barad's ontological framework of this intra-active process, where bodies simultaneously produce and are produced by each other cannot be over-emphasised. Fundamentally, it is Barad's contention that everything, including humans and non-human creatures, quantum superpositions and earth systems, music and language, are all entanglements and diffractions, rather than separate and permanent formations. The consequence of this is that we never stop becoming entangled with the world that we inhabit; never stop producing and being produced by everything that is around us: a signally ecological and interdependent way of understanding the world.

Before completing this exploration of Karen Barad's work, it is also important that we further examine how interiorised exteriority operates at the level of human experience. Just as Meillassoux's lasting contribution to philosophical discourse may be the idea of correlationism, then Barad's concept 'agential realism', may be their legacy. Agential realism describes how exteriority (that is, a non-correlated, non-subjective real) is produced by an individual (whose experience of being in the world is located within a world that they can only experience subjectively) making what they refer to as an 'agential cut' (ibid.: 348). We can think of this cut as a moment of observation and, as we shall see in the next chapter, elsewhere in *Meeting the Universe Halfway*, Barad also likens the process of observation to measurement and mark-making. We might also consider its relation to decision-making. However we choose to conceptualise the agential cut, the essential point that Barad makes is that, within the process of intra-action, exteriority is produced by being absolutely conditioned by its time and place of emergence. In short, agential realism describes the process of a non-subjective, non-correlated real being produced via an individual's subjective observation of the world. What is more, Barad describes the agential cut as a 'contingent resolution' (ibid.: 348), indicating – as does Meillassoux – that there is no underlying reason for things being the way they are; any other outcome is also always possible.

New Materialism: Vibrant Matter and Naturecultures

Another theorist whose work is seen as part of the new materialist project is Jane Bennett. In *Vibrant Matter: A Political Ecology of Things*, Bennett uses the concept of 'vibrant matter' to convey the agency of seemingly inanimate matter. Bennett describes electrical grids, power blackouts, large ecosystems and the planet earth itself as examples of vibrant matter, whose 'parts are both intimately interconnected and highly conflictual' (Bennett, 2010: 23). In describing the effects of a power blackout, Bennett describes how human intention may be folded into the workings of a greater whole, where there is 'not so much a doer (an agent) behind the deed (the blackout), as a doing and an effecting by a human-nonhuman assemblage' (ibid.: 28). Bennett is initially careful to maintain a distinction between assemblages and organisms, using an organism's ability to self-organise in order to distinguish between 'humans and their (social, legal, linguistic) constructions [and] very active and powerful nonhumans: electrons, trees, wind, fire, electromagnetic fields' (ibid.: 24). As her analysis deepens, however, she begins to explore the question of agency that the power blackout raised, and she asks whether an assemblage, like an organism, has the capacity to self-organise and form a culture (ibid.: 34). As a result, she proposes that although human agency is normally held in higher regard than material agency, or even material–human interaction, humans themselves are in fact assemblages of nonhuman agents, and therefore are not fundamentally different from nonhuman assemblages. For Bennett, it is a difference in scale, rather than a difference in kind:

> On close-enough inspection, the productive power that has engendered an effect will turn out to be a confederacy, and the human actants within it will themselves turn out to be confederations of tools, microbes, minerals, sounds, and other 'foreign' materialities. Human intentionality can emerge as agentic only by way of such a distribution.
>
> *(ibid.: 36)*

Here, Bennett moves the focus from the agency of the living seen as autonomous, self-propelling phenomena, and suggests that humans, as assemblages of tools, microbes, minerals and sounds, are constituted as much by non-living as by living elements. Bennett's claim that the interactions of non-living matter and energy are the driving forces behind events, such as power blackouts, suggests that being alive is in fact just an ephemeral configuration of otherwise inorganic material. In other words, as far as Bennett is concerned, inanimate, non-sentient things have agency in the same way that sentient beings have agency, again: all things are essentially the same type of thing.

As we have seen, Barad's ontology is not as open as this, and does differentiate between sentient and non-sentient things. Indeed, observation – in the form of subjective experience, decision-making, creative production and other interpretations of quantum measurement – is at the heart of Barad's model, unlike Bennett's conception of agency, which is a more generalised capacity to engender events. Similarly, in an agential realist epistemology, thinking and touching have no greater access to the real than the other, discourse is no more or less material than physical presence. The signal difference between Barad's and Bennett's models is that, although human experience when seen as entangled intra-action might be operating in the same way as anything else in the world, agential realism folds in subjective experience as part of its mode of operation, thereby fusing subjective experience of the world with the capacity to create the objectively existing world. Moreover, though we are connected to and part of the world around us, agential realism produces a cut that fundamentally separates from us from the things around us. Thus, Barad, as does Meillassoux, offers a mode of thinking that allows for interiorisation of exteriority that enables us to move on from Naess' closed loop of abstract structures.

In their essay, 'The Transversality of New Materialism', Rick Dolphijn and Iris van der Tuin attribute the emergence of new materialism – and 'neo-materialism' – to the work of Manuel DeLanda and Rosi Braidotti in the 1990s (Dolphijn and van der Tuin, 2012: 93), describing it as 'a cultural theory that is non-foundationalist yet non-relativist' (ibid.: 110). Such a formulation echoes Meillassoux's vision of an anti-correlationist speculative realism which, as we have seen, also maps an epistemology that encompasses – but is not founded on – an absolute real that simultaneously exists outside of the limits of human thought, while remaining conceivable by human thought. In their 'cartography' of new materialism, Dolphijn and van der Tuin recognise the role that the multi-species feminist, Donna Haraway, played in mapping out the territory that the likes of Barad, Bennett, Braidotti and others would go on to explore. Although not a 'new materialist' in the explicit sense articulated by either Dolphijn and van der Tuin, or Coole and Frost in their collections, Haraway's work has been a key foundation for new materialist thought, and indeed has directly influenced the work of Karen Barad. Dolphijn and van der Tuin cite Haraway's use of the term 'material-semiotic' which describes how objects of knowledge come into being via their interactions with other 'generative nodes' (ibid.: 109), a concept that we also find echoed in Levi Bryant's focus on material forces and human relations that we encountered in the Introduction. Since the interactions between these nodes create boundaries, which themselves become sites of signification, for Haraway, material interaction is therefore a process of signification and meaning making. Indeed, Barad has acknowledged how Haraway's sense that 'diffraction patterns record the history of interaction, interference, reinforcement, difference' (Barad, in Dolphijn and van der Tuin, 2012: 51) underpinned their own extensive use of diffraction as a means to describe the inter-/intra-active process of meaning- and mark-making that resides at the heart of *Meeting the Universe Halfway*. In this sense, it becomes clear that even the title of the book is a direct reference to how – after Haraway – 'diffraction is about heterogeneous history, not about originals' (ibid.: 51), where meaning and the real emerges via the process of ongoing collision and reformulation, both of matter and of discourse, which – following Naess – is itself an ethical position that affirms an enhanced ecological egalitarianism. The Barad–Haraway conjunction here is the sense in which intra-action is the process of meaning, and the experience of the real, coming into existence, as a located force that produces not only meaning, but the site for the possibility for further collisions, and further meaning-making.

Haraway herself draws on the work of evolutionary biologist Lynn Margulis, using the term 'holobiont' to describe how humans – as 'symbiotic assemblages [...] are like knots of intra-active relating in dynamic complex systems' (Haraway, 2016: 60).[3] As a biologist, Haraway's work is full of references to biological formulations of the world, and it is worth noting in this instance that her views about intra-action and complex systems are very much aligned with a multi-species view of how the world functions. Although we humans exist as autonomous entities, we also exist in symbiotic relationships with a host of microbiota. Indeed, the 'occupied habitats' of the human microbiome include our 'oral cavity, genital organs, respiratory tract, skin and gastrointestinal system' (Kho and Lal, 2018), and while we host and nourish these microbiota, we also survive by virtue of their presence in our systems. Haraway engages with the two poles of this relationship, and contrasts autopoietic processes as self-producing systems – or 'autonomous units' – with sympoietic – 'collectively-producing' – systems (Haraway, 2016: 61).

Although she draws a distinction between autopoietic and sympoietic processes, Haraway sees the two terms as operating together – in a process she likens to 'generative friction [...] or enfolding' – as different aspects of complex systems (ibid.: 61). Haraway also uses Margulis' term 'symbiogenesis' to convey how humans, as all living things, are deeply involved with each other in the production of new things, new patterns and new formations – an approach that '[ties] together human and nonhuman ecologies, evolution, development, history, affects, performances, technologies, and more' (ibid.: 63). In this regard, Haraway's work is both a precursor to and a corollary of the new materialist project, providing us with further insights into the patterning processes that constitute the production of meaning and experience, and at the same time giving further emphasis to the necessary intertwining of ontology, epistemology and ethics. As Dolphijn and van der Tuin suggest, where Barad referred to 'an onto-epistemology, or even an ethico-onto-epistemology, according to which being and knowing (and the good) become indistinguishable', their work drew heavily on Haraway's conceptions of multi-species world making ('worlding') and intra-active patterning (Dolphijn and van der Tuin, 2012: 110).

Non-philosophy

Where the speculative realists argued that human thought must be able to access something which is fundamentally 'outside' itself another continental philosopher, François Laruelle, whose career spans the early 1970s up to the present day, had also been engaged in a similarly radical project to develop the idea of 'non-philosophy'.[4] Via a sustained exploration and dismantling of what he described as 'philosophies of difference' (Laruelle, 2010b), namely the work of Nietzsche, the phenomenology of Heidegger and the post-structuralist contributions of Derrida and Deleuze, Laruelle developed the concept of non-philosophy as a means to rethink our understanding of what philosophy is and what it can achieve. Although Laruelle's non-philosophical project is deliberately designed to set his practice outside or against standard philosophical enquiry, in *Laruelle: Against the Digital*, Alexander Galloway maintains that Laruelle is a materialist thinker 'in a very basic sense' (Galloway, 2014: xxviii), going on to describe Laruelle's approach as a 'weird materialism' (ibid.: 8). Galloway lists Laruelle's commitment to finite materiality and lived experience as exemplified by the 'real material experience of human life', along with his direct references to Marxist thought and historical materialism, in particular the relations between forces of production and relations of production, and the determination of the material base (ibid.: xxviii). As such, examining some of the core features of Laruelle's work enables us to develop a wider perspective on materialist tendencies in twenty-first-century thought, and how this might find expression in our focus on music and its related practices in later chapters.

In the book *A Non-Philosophical Theory of Nature*, the theorist and Laruelle translator Anthony Paul Smith provides us with an overview of Laruelle's decades-long non-philosophical project and enables us to understand the way in which, for Laruelle, the practice of philosophy has no greater claim than science, art, or indeed music, as a means to engage with what Laruelle defines as 'the Real' (Smith, 2013). In the context of non-philosophy, the Real is the term that Laruelle uses to describe that which lies beyond human categorisation and differentiation (hence his sustained dismantling of the above-mentioned philosophies of difference), but which nevertheless gives rise to everything that we experience.[5] One of Laruelle's key ideas and claims is that conventional philosophical thought places itself above other modes of human enquiry and thought, via what he calls the 'Principal of Sufficient Philosophy' (Laruelle, 2013a: xxvi). The inference here has been that practices such as science, mathematics, music and art are merely conventions and have no access to reality in itself. Only philosophy can make a claim to be able to describe or engage with the world as it is in itself, or with the Real. However, for Laruelle, philosophy and science are doing the same kind of thing: both are practices that are given in the Real – which means that each one operates in the world of real, human experience that is produced within the absolute Real – and, because of this, neither has any more or less privileged access to the Real. This is why Laruelle makes use of the prefix 'non' in non-philosophy and non-standard philosophy: it tells us that we need to reconsider what it is that we think philosophy is doing, and what it – and any other form of human practice – is able to do. Therefore, whichever term we use, it is important to understand that Laruelle's project is essentially a recalibration of the scope and limits of philosophy; correcting an over-reaching historical precedent – one that sees philosophy as having a privileged relation with matters of ontology and epistemology – by rebalancing it alongside other practices of knowledge creation.

As we have seen, Laruelle uses the term 'the Real' to denote a ground for experience which in itself defies experience, a position that resonates with our earlier discussions of the absolute, and Meillassoux's 'great outdoors': we can know that a ground for everything that we experience exists, but we cannot know anything about it. While Laruelle's Real similarly refuses representation and exceeds human perception, one of the more remarkable things about his approach is that he sets out to design a method for thinking *from* this Real, instead of simply thinking about it. Non-philosophy is a way of thinking outwards from an immanent absolute, rather than thinking about how such an absolute might interface with the world of human experience, and Laruelle uses the figure of 'the One' to denote this absolute Real. In *Philosophy and Non-Philosophy*, he proposes that we can think of the One as being alongside us, as something that is immediately given to us: it is 'what we are intrinsically in our essence' (Laruelle, 2013b: 45). This is to say that everything about us, as humans in a real world of lived experience, is simultaneously part of the inaccessible Real of the One. Brassier explores the distinction between these two modes of the R/real. On his analysis, although the human subject of Laruelle's non-philosophical model can differentiate between the immanent One that is the 'determining essence' of everything, and their own power of 'decision' which is 'determinable' (Brassier, 2003: 30), the One itself is 'indifferent to [human] decision' and remains indeterminable, that is, foreclosed to human thought. We therefore have a set of vectors within Laruelle's model that we can map out. The One-Real is the all-encompassing frame which envelops and determines everything that can be experienced. We humans make decisions within the context of the real world of lived experience, and we are able to articulate our own reasoning processes that underpin our decisions. Similarly, we are able to recognise that the One-Real is the determining factor within which the real world is given, but we have no capacity to interrogate or engage with what it is or how it works.

Although complex, Laruelle's concepts of the One-Real can be set out in a relatively straightforward manner: the One is immanent to itself without cause or necessity, and is radically immanent to everything. In other words, everything is 'given' in the One. This is the sense in which Laruelle's rationale for the non-possibility of a Principle of Sufficient Philosophy works: only the One can be immanent to itself and self-sufficient, and therefore philosophy is itself simply given in the One (Laruelle, 2012c: 394). Laruelle uses the phrase 'vision-in-One' to communicate his sense of what human experience is: it exists at the level of the real, but is afforded, or determined by the Real/the One. In *Philosophy and Non-Philosophy*, Laruelle describes vision-in-One as 'the experience of this real' (Laruelle, 2013b: 54). Laruelle frames what this experience might be 'like', suggesting that it is a reflection of the Real, which, as we shall see, relates to his ideas about cloning. While we may not be able to 'know' the Real, Laruelle's non-philosophical approach proposes that we are able to 'perform' it through creative acts of production thinking, decision-making. Laruelle uses the concept of 'cloning' the Real to describe what happens in the creative act of producing the world. As the word suggests, cloning is not an actual engagement with the Real, it is more like a process of copying it. The mirror analogy holds well here, in that any creative act is a reflection of the Real (as opposed to a representation of the Real). Laruelle's indication that the Real is both alongside us, and at the same time the essence of what we are, is also useful. A creative act – taking a photograph, making music – is both essentially Real, and a clone of the Real that we are able to experience in the real world. The image of the clone communicates well the double aspect that human-made things have. We can both experience them, and not experience them at the same time. Music is both determinable, because someone has made it, and indeterminable, because it is essentially comprised of the indeterminable Real. In a literal sense, vision-in-One frames human experience (vision) of the world it is able to experience (determine), as something that is afforded by the One (as the fundamental cause that humans cannot experience). Laruelle uses the phrase 'unilateral causal relationship' to further clarify this situation. This causal relationship can only ever move in one direction from the One; there can be no reciprocal engagement between thought and the One, and thus the One cannot be affected by thought. This unilateral, one-way causality is what Laruelle means by the phrase 'given-in' that occurs repeatedly throughout his work, and it is used to convey the sense in which everything that has been caused by the Real (which for Laruelle is everything that there is) has no subsequent access to its own point of origin. Everything within Laruelle's non-philosophical framework centres on this idea of unilaterality, where it makes no sense to speak of any kind of 'access' to the One. Hence, the vision-in-One arises in the One in the same way that thought arises in the One, and as we saw above, human investigative and creative practices such as science, philosophy and music are all therefore 'given-in-One'.

The vision-in-One is not a description of experience as such; instead it is an indication of how we can refer to the experience of being 'given-in-One'. Thus, vision-in-One is a statement about how we come to understand the world as a context that we are given into, and how this shapes the context and horizon of our knowledge. Moreover, it is essentially why Laruelle repositions philosophy as a practice of knowledge alongside science. Philosophy, science and practices of making are all modes of engaging with, and creating, the world that are given-in-One, and generate vision-in-One. The active non-philosophical principle is that any human activity is R/real simply by virtue of its being given in the One; nothing has any privileged status in terms of its capacity to provide access to the One-Real. Finally, vision-in-One is also a further reflection of Toscano and Kinkle's statement about the non-representability of the absolute: the vision-in-One does not 'represent' a relationship with the absolute, it is the process of being in the absolute. Unlike Meillassoux's great outdoors, Laruelle's model does not allow for any knowledge of the absolute reality that the One-Real exemplifies. Where Meillassoux developed a formulation that allows us to produce a tangible knowledge about the absence of knowledge, Laruelle adamantly denies any such process. For Laruelle, the outcome is both far simpler and at the same time equally challenging: vision-in-One means that everything in the world is unavoidably Real, but we can never grasp or understand what this reality is or what it means.

Although Laruelle's argument is complex, it is possible to understand the fundamental point that everything is given within the One, a variation on a theme that we have now encountered three times. The speculative realist position suggests that the world of human experience is conditioned by an absolute and necessary contingency, Meillassoux's proposal for a non-conceptualisable Real. The absolute, objective truth about the world we inhabit is that, beyond the fact that it exists, we really are not able to say anything about it that is in itself 'true' or 'real'. This truth about the unknowable nature of reality resides in the worlds that we experience, it is a truth about the Real that is presented in the world, but a truth that we cannot determine. Indeed, Alexander Galloway discerns a commonality between Laruelle and Meillassoux in their shared antipathy towards the function of human experience in the process of apprehending the real, suggesting that for each of them, 'the human is considered to be a limiting factor vis-à-vis what may be known about the real' (Galloway, 2014: xxi). Although the agential cut is a means of allowing for a subjectively located creation of objective reality, Barad's new materialist perspectives on observation are designed to show that absolute determination of the world is categorically impossible. However, the concept of intra-action, and its connection to the way that agential cuts produce simultaneously located subjective-objective 'marks', that momentarily dis-locate the shifting, topological flow of intra-activity, is akin to Laruelle's concept of the vision-in-One. Finally, while deep ecology has the most tenuous link to non-philosophy by some margin, nevertheless Naess' quantum physics-informed material ontology lays out a vision for a flat world of experience that is made up of the concrete content that humans produce, but are not able to determine. In simple terms, deep ecology expresses a view that humans are able to produce a R/real that they are not able to understand, and that abstract structures become a way of labelling that R/real; describing it, but not determining it.

Objections

In the years since *Meeting the Universe Halfway* was published, much has been written about new materialism, including scholarship that has both sought to extend and critique its key themes. Along with Coole and Frost's work (2010), Dolphijn and van der Tuin edited a collection of 'Interviews & Cartographies' (2012), including interviews with both Barad and Meillassoux. Subsequently, Andreas Malm (2018) and Paul Rekret (2016, 2018) have each voiced concerns that – as part of the 'new' materialism movement – Barad's conceptual developments have undermined the ethical foundation of Marx's original historical materialism project. Robin James (2019) criticises Barad's work on the grounds that – as with other new materialists – it draws too heavily on abstractions – in Barad's case relating to quantum physics – to the detriment of engaging with actual people and their lived experience.

Rekret sees the new materialist project as espousing little but 'jargon', suggesting that these 'theories of matter' only work because they deliberately screen-out any sense that the human subject under consideration is 'formed and deformed' by the objective mechanisms of the 'social logics and institutions of capitalism' (Rekret, 2016), describing both Barad's and Bennett's work as 'naïve empiricism' which operates on the basis that there is a 'continuity between ontology and ethics' (Rekret, 2016). He is not necessarily wrong to highlight such a continuity, since – as we have seen – Naess also identified a mediating role that abstract structures played between subjective experience and the world. However, Rekret frames social relations slightly differently from Naess, seeing capitalism as an objective structure that creates social relations which then function in a particular mode. In 'The Head, the Hand, and Matter' (2018), Rekret goes on to argue that the current separation of thought from matter is the legacy of radical changes in social relations, caused by industrial capitalism, in the shift from agricultural, artisanal work to an industrialised organisation of labour. In this more recent essay, his concern is that the new materialists' 'failure' to recognise that social relations fundamentally underpin a conceptual environment that either separates or reconciles human and nonhuman worlds has 'pernicious effects upon their political analyses' (Rekret, 2018). Essentially, as far as Rekret is concerned, new materialism is a jargon-infested, ethical vacuum, that simply entrenches the political imperatives of capitalism by not acknowledging the political-economic forces at work that created the change in social relations between humans and the world, and which continue to mediate our experience with the world. Rekret's central concern rests on his sense that 'our relation to the world and to nature are themselves the result of collective struggles' (Rekret, 2018), and that the mode in which the new materialists operate completely occludes the contingent, and politically underpinned sociopolitical environments that we inhabit. In essence, the way that we think about the world is completely the result of human struggle and political imperatives. Rekret's work thereby pushes back against both the Material and Speculative Turns. For Rekret, agential realism and its kin may have broken us out of the correlation, but at the cost of acknowledging the mediated nature of the world.

However, where Meillassoux's work operates in a highly abstract mode, and often gives little indication that he has considered any kind of practical dimension to how the results of his insistence on the existence of an outside to human thought might manifest within thought – or what the consequences of this might be in a practical sense – Barad's work does indeed offer more purchase. Indeed, the equivalence they draw between discourse and matter, folded as they are into the intra-active model, suggests that social relations are indeed an essential part of how we negotiate the world, and are produced by the world that we inhabit. In this sense, whereas Meillassoux's world may have the appearances of a world free of context and the social relations that have produced the modes of thought we currently use, Barad's work gives primacy to the forces of social production, even if they are not named as such.

Rekret reflects on how 'non-human natures and technologies have entered into social relations in new ways in recent decades', but expresses his concern that 'the possibility of reconciliation of human and nature, or thought and world, cannot be articulated in terms of ethical experience' (Rekret, 2016). This suggests that the new materialist ideas espoused by the likes of Barad and Bennett lack an ethical dimension that sufficiently reconciles humans with 'non-human natures', beyond nature beyond simply a commodifiable resource. In other words, while new materialism tries to create a holistic picture of humans as part of the world, its ethics are too simplistic and human-centred, which for Rekret is not any kind of whole picture at all. Is there necessarily a necessary ethics though? While Rekret is right to observe that the industrial revolution created a rift in relations between human and non-human natures unlike anything that had occurred up until that point, was the pre-industrial consensus any more ethical – inherently or otherwise – than what came afterwards? Of course, ethics are a concern, but are ethics produced by eternal verities, or – as Barad might suggest – do they result from agential cuts and diffractive overlays and accretions over time? As both Barad and Meillassoux show, ethics are 'real', but this isn't to say that they are either necessary or eternal.

While Rekret may be entirely correct to question the new materialist project on the grounds that it undermines – or at least problematises traditional notions of materially located politics – this is not an end to the matter. What he does not acknowledge is that Barad continually emphasises that our understanding of what matter is has fundamentally shifted since the original formulation of these original materialist perspectives. The latter's commitment to exploring and generating a coherent account for the entangled nature of matter suggests that conventional modes of apprehending and accounting for the nature of matter may be in need of radical updating. This would suggest that Rekret's concerns about the de-politicising tendencies of new materialism may yet prove to be unfounded, since what in fact may be required is a new theory of material politics to accompany the new materialist vision.

Andreas Malm, in 'The Progress of this Storm' – described as a 'furious defence of dialectical thought' (Malm, 2018) – attempts a similar dismissal of Barad's work. He criticises their apparent quest to equate non-human forces with 'characteristics of intentional or purposive behaviour', and to argue that agency and intentionality are not uniquely human matters. However, there is little evidence in Barad's work of them suggesting that subatomic particles display 'intentional characteristics', and neither do they decouple agency from 'any kind of subject at all' (Malm, 2018). On the contrary, *Meeting the Universe Halfway* continually emphasises the centrality of the human subject in the process of making agential cuts, and agential realism is a project designed to validate the existence of an objective reality within subjective human experience. Reza Negarestani has levelled similar criticisms, claiming that 'the credo of posthumanism, critical vitalism, and new materialism, strives to remove any reference to human exceptionalism in any sense, and instead replaces it with natural objects, processes, and material networks that can undermine human exceptionality bottom-up' (Negarestani, 2020: 127). Again, such a 'bottom-up' approach that Negarestani and Malm alight on does not, in fact, hold all the way up. Though Barad starts by positing the material commonality of the world, this is not the same as saying that humans and particles are the same type of thing. For Negarestani, new materialism takes the form of 'a thorough-going deconstruction of the concept of the human by material networks and actants (posthuman ontology from below) of which the human is the only one example' (ibid.: 127). The issue, as he sees it, is that this process of deconstruction 'overlook[s] what is most significant about the human', thereby rendering it obsolete (ibid.: 127). In essence, what Negarestani is accusing Barad and their new materialist colleagues of doing is not only reducing the human to its parts, but insisting that the human is the same as its parts. It is far from clear that this is the line that Karen Barad is taking with their work. Indeed, to an extent, their position resonates closely with Negarestani's description of the human as a work 'under construction' (ibid.: 128). We shall explore this aspect of Negarestani's work in Chapter 5, but to note briefly, it is a position that suggests that humanity, far from being a complete and resolved entity with qualities that are essentially 'human', is instead a work in progress; a specific type of entity that is evolving and changing through history. Barad's quantum-inflected realism presents an ontology where things exist within, and are simultaneously constructing, a topology; the agential cut-as-observation is simply a reading of the human at a given point. Haraway's work is also relevant here. She uses the term sympoiesis – 'making-with' – to convey the idea that 'nothing makes itself; nothing is really autopoietic or self-organising' (Haraway, 2016: 58). As with Barad's conviction that the world – including humans – is fundamentally a process-in-motion, sympoiesis enables us to reflect on Negarestani's sense of human exceptionalism and his assertion that the human is a point of self-generation and self-determination. Certainly, the human is a point within a topology that produces a real point of experience, but this is a process that the human shares with everything around it. The human is not self-produced, and the human-under-construction is not simply constructing itself. It is, however, part of an ongoing topology that is continually being remade and re-making itself.

Along with Rekret and Malm, Negarestani's criticisms are also levelled at Jane Bennett's work. While Bennett's conception of vibrant matter is less resolved than Barad's agential realism, and also lacks the extensive grounding that quantum physics provides in terms of the technical details of how agential realism works as both an ontological and epistemological model, nonetheless Negarestani's perspectives are not entirely without their own discontinuities. Rather than simply equating humans with microbes, electricity and thunderstorms, as a concept, vibrant matter seemed to suggest that the way that thunderstorms and humans catalyse events in the world is not as fundamentally different as we might think. 'Agency' is the word that Bennett uses to describe a non-essential difference that interests her between humans, thunderstorms, earthquakes and microbes, again, gesturing towards defining the difference as a difference in scale, rather than in kind. Clearly, Negarestani is committed to maintaining a sense of human exceptionalism; that humans are a very particular type of entity in the world, but again, Bennett is not necessarily proposing that everything is made of the same materials, displays similar properties, and is therefore the same. Nonetheless, Negarestani's critique is a sharp reminder to remain vigilant, and not be swept along with a flat ontological reading of the world that simply reduces everything to a featureless plane.

In *The Sonic Episteme*, James' analysis of Barad's work alights on what she sees as the new materialists' predilection for a reliance on generic, abstract concepts over specific, and located, analysis. Locating the wellspring of new materialism in white male academia, James' focus is to identify a hidden exceptionalism in new materialism's flattening of everything into one plane of discourse. Thus, on the one hand, where Rekret pushes back against Barad et al.'s erasure, or occlusion of an epistemological framework that shows how this new mode of materialist thinking operates, on the other, James resists its claims to universalism on the grounds that – as with any other school of thought – it is a mode of thinking that is produced in a certain way, by certain groups and with a certain goal. Thus, for James, new materialism's supposed neutral abstraction is a disingenuous, non-self-critical mode of thinking, that masks its exceptionalism in its pretensions of universalism. Amongst her criticisms of Barad's work is the latter's 'failure to engage black studies and work by black scholars' (James, 2019: 114). In this respect, James puts forward a series of what she describes as 'nonelite' alternatives to the learning that Barad has taken from quantum physics, including Christina Sharpe's theory of 'wake' and Ashon Crawley's concept of 'choreosonic vibration' (ibid.: 110–20). While it is true that Barad does focus on general rules relating to the behaviour of particles at the quantum level, rather than on movement of specific particles, their concept of agential realism is in many ways a strong analogy for how James understands lived experience to be a valid source of alternative philosophical and epistemological models.

Given these objections to new materialism on the grounds of its political vapidity,[6] it is interesting to note that Alexander Galloway reads new materialism as a response to a particular set of politically motivated critical trajectories. He suggests that 'we need to acknowledge that this current round of realism is in fact a direct response to – and a desire to do away with – projects that were themselves extremely political, projects like identity politics' (Galloway, 2010: 23). In that sense, while the new materialist milieu takes on a decidedly apolitical hue in its disavowal of politically inflected philosophy, it remains a philosophy focused on creating a new politics, one that is deliberately couched within a new materialist environment. Indeed, on the question of ethics, Galloway likens new materialism's rejection of structuralism, post-structuralism and other theoretical models conditioned by critical thought, to phenomenology's rejection of scientific positivism in the early twentieth century. Asking 'What does phenomenology tell us?', Galloway suggests that the message of phenomenology was that 'all philosophy should start, not from the real, but from the life world; it says you should start from someplace other than math or positivistic science. There's a kind of morality in Husserl's position' (ibid.: 23–4). It is certainly noteworthy therefore, that while Rekret, Malm and James decry new materialism for having expunged philosophy's ethical core, Galloway celebrates its ethical move – in a manner not unlike Naess' articulation of deep ecology's ethical imperatives – that new materialism is an ethical project by virtue of what it is designed to do, and that by focusing on the methods of its operation in the way that James and Rekret do, we miss its overarching trajectory as a model of thought. In asking, 'What does materialism tell us?' Galloway informs us that

> it tells us that everything should be rooted in material life and history, not in abstraction, universality, logic, necessity, essence, pure form, spirit, idea, etc. I see it as similar to the Foucauldian question about power. It's not that power is good, or power is bad – power is dangerous.
>
> *(ibid.: 23–4)*

This suggests that not only does Galloway see new materialism as an ethical project that is designed to create a materially inflected discourse to engage with the world, one that is rooted in lived experience, but also that – far from being an abstract, vapid and apolitical philosophical project, it is a highly charged and unpredictable means of approaching how we understand the world.

As we have seen with all of the speculative realist and new materialist thinkers in this chapter who fundamentally position any mode of discourse in the world, such as language and ethics, these discourses are constituted by everything else in the world, and they simultaneously constitute everything else in the world too. What unites these thinkers is their suspicion of some form of correlationism or philosophical idealism, where reality is a function of the mind, or as we see in Meillassoux's work, the world is a product of our capacity to know it. Mathematics, ethics, capitalism, speculative realism, humans and new materialism; none of these are neutral, and all of them have a huge part to play in how the others are constituted and operate.

Conclusion

Arne Naess' ideas facilitated the development of an ethical materialism that centred on how a recognition of the interdependent nature of the world led to an empathetically driven epistemology. In the context of the speculative realist, new materialist and non-philosophical perspectives explored in this chapter, a speculative materialist ecology comes into view, one that is ethical because it affirms that events really do happen, perspectives really do come into existence, and meaning really can have an impact on the world.[7] In this sense, nothing is just an opinion; everything really is an agent of change in the real world. Negarestani's insistence on human exceptionalism and Barad's agential realism can exist within the same frame; human exceptionalism is one amongst many really-existing, unique events in the world.

In one sense, we are fundamentally different to, and disconnected from the world, and in another, we are completely connected to the world; able to make marks on the world that are objectively real differences and changes. Music is one such mark. To an extent, Laruelle's non-philosophy addresses both of these vectors. His ideas about the Real and the One reflect the new materialist and deep ecology view that, at a fundamental level, there is simply one interrelated reality. This One-Real is experienced at the level of the individual as the real, or the world, and consists of sensory experience and the subjective experience of being an individual. However, it is not that subjective experience allows for the experience of the world, it is that individual experience of the real is a clone of the Real, and is a function, or an aspect of the One-Real. The key component of non-philosophy that differentiates it from deep ecology, and certain aspects of new materialism is that human processes and activities are both an intervention into the real world, but also a channelling, or rather a cloning of the Real. This is to say that when we make a piece of music, we are creating something that becomes part of the real world that we exist in, but we are also operating in the way that the Real-One operates, and creating the Real, creating the absolute. We cannot access or even properly conceive as such what this Real is, but nevertheless, we are able to create it. This is the similarity with speculative realism, non-philosophy's proposal that the Real really exists, separate from our capacity to imagine it, and both approaches allow for a conception of a reality that is not simply a product of our minds. The speculative realists' outside and Laruelle's One-Real are fundamentally beyond comprehension, and we are not able to make any kind of coherent statement about what these absolutes might be.

The challenge of this chapter has been to assemble a range of sometimes contradictory perspectives that will enable us to think about music, and creative practices of making, listening to, sharing and monetising music in a way that enables us to recognise the material relations between these practices, and the ecological nature of these relations which means that all of these changing and interchangeable relations fundamentally shape each other. As we shall see in the following chapters, 'mattering music' is a way of apprehending music's real, material relations with technological, economic, creative, as well as social and natural environments, and this means that, following Barad, Meillassoux and Laruelle, music is both produced by, and produces, each of these worlds.

Notes

1 Post-structuralism was a mid-to-late twentieth-century movement in critical thought which, in essence, problematised the notion that meaning is stable and fixed, proposing instead that meaning is constantly in flux and shaped by power relations.

2 Since the early twentieth century, analytic philosophy has emphasised the use of logical and linguistic analysis to better understand concepts and problems. Analytic theories are grounded in empirical evidence, formal logic and mathematics, and foreground the importance of precision in language.

3 Lynn Margulis collaborated with James Lovelock on the development of the Gaia hypothesis in the early 1970s, a theory that was designed to show that from its earliest emergence 3.7 bn years ago, life on earth has controlled the planet's environment via a process of homeostasis that regulates the biosphere (Margulis and Lovelock, 1974).

4 In his later works, for example *Philosophie Non-Standard: Generique, Quantique, Philo-Fiction* (2010c), Laruelle has come to refer to non-philosophy as 'non-standard philosophy'. Largely, this has been seen as a move to clarify that his approach is not against philosophy as such, simply that he is interested in repositioning the objectives and limits of what has become standard philosophical practice.

5 Throughout this chapter, we can make a distinction between two registers of the real. The Real with a capital 'R' – which is interchangeable with 'the One' – denotes the absolute reality within which everything exists, but which we have no means of accessing or conceptualising. In this sense, the Real bears a relation to Meillassoux's conception of the 'outside' of thought. The real with a lowercase 'r' – as the real which is given in the One – denotes the real experiences and thoughts of Laruelle's 'lived' world, a real that we come into contact with, and which we knowingly interact with. The lowercase real is 'real' by virtue of the fact that it is given in the Real/the One; since anything that is given in the One is real (because it is of the Real).

6 Given the strength of this Marxist critique of new materialism, it is important to note that, in a broad sense, one of the key reference points for Laruelle's use of the Real as a foundational context in which the world of experience is given, is Karl Marx's base-superstructure model, as outlined in the Preface to *A Contribution to the Critique of Political Economy* (Marx, 1904: 11). Marx's model, which places emphasis on the 'totality of relations' and the idea that humans enter into these relations independently of their free will (ibid.: 11), is a clear forebear of the One as all-encompassing framework that we see in Laruelle's thought. By establishing that everything is already within a non-philosophical Real-One, Laruelle parallels the relations of production as the 'real foundation' of social consciousness, by suggesting that the One is just such a foundation for human consciousness. However, this sense of being given within a real foundation is not the only idea that Laruelle borrows from Marxist thought, and in his introduction to *Future Christ* (Laruelle, 2010a), Anthony Paul Smith sketches a definition for another key non-philosophical concept, the notion of determination-in-the-last-instance, Laruelle's adaptation of a key component of base-superstructure theory. In *For Marx*, Althusser references Friedrich Engels' discussion of the relationship between the productive foundation of society and its political, social and cultural superstructure, to establish that, 'production is the determinant factor, but only "in the last instance"' (Althusser, 1969: 111). Althusser is suggesting that, for Engels, it was vital to establish that the relationship between the base and the superstructure remained entirely reciprocal, and that the one-way process wherein the superstructure is solely determined by the relations of production would only occur when no other forces were in play. In *From Decision to Heresy*, a collection of Laruelle's writing from 1985 to 2012, Laruelle usefully provides us with his own reading of the 'last instance', which as we can see from his brief definition, is fundamentally another way of describing the One: 'By last instance, I describe that which is real in itself, that is to say that which has no need of existence in order to be real. Or that which the description as real in itself has no need of this description in order to be real in itself, and of which it must be constituted' (Laruelle, 2012c: 395).

In one sense, determination-in-the-last-instance is another reference to the process that we have already encountered wherein everything in the real world is given in the One, but it is also a means of capturing Laruelle's modification of Engels' statement, which suggests that the real is always unilaterally determined by the One-Real. The concept of 'determination' also enables Laruelle to talk about causality in a very particular way; and where we saw above that the vision-in-One is a tracing of a non-philosophical causality, determination allows us to understand that although the real world is given in the One, and is therefore determined by the One, this does not mean that it is directly 'caused' by the One. We cannot say 'how' things come into the world, because we can have no knowledge about the One, we can simply say that they are in the One. As Smith suggests, since we are unable to 'think the Real in any meaningful sense [...] the point of non-philosophy is simply not to think the Real' (Smith, 2013: 69), which is to say that non-philosophy puts the Real to one side as it were, and instead directs our attention towards the process of thinking 'from it' (ibid.).

7 In *Realism Materialism Art*, Christoph Cox, Jenny Jaskey and Suhail Malik examine the relationship between of materialist and realist philosophical perspectives. Where materialists emphasise the material nature of the world, particularly in terms of material forces and material processes, realists take the view that reality is independent of the human mind. In this sense, materialists are often realists. However, since for realists symbolic meaning is real, the reverse is not necessarily the case (Cox, Jaskey and Malik, 2015: 25). With regard to the work of Laruelle, Morton and Meillassoux, Cox, Jaskey and Malik see that the non-philosophy project operates in a realist mode and rejects materialism, as does Graham Harman's 'Object-Oriented Ontology', which Morton has drawn from extensively across a series of books. As for Meillassoux, on their reading, he plays down the speculative realist trajectories in his work, opting instead for what he terms 'speculative materialism', so as to differentiate the realist aspects of his approach from naïve realist perspectives on reality.

Bibliography

Althusser, L. 1969. *For Marx*. London and New York: Verso.

Avanessian, A. and Malik, S. 2016. *Genealogies of Speculation: Materialism and Subjectivity since Structuralism*. London and New York: Bloomsbury Academic.

Barad, K. 2007. *Meeting the Universe Halfway*. Durham and London: Duke University Press.

Bennett, J. 2010. *Vibrant Matter: A Political Ecology of Things*. Durham and London: Duke University Press.

Brassier, R. 2003. 'Axiomatic heresy: The non-philosophy of François Laruelle', *Radical Philosophy*, 121, September/October. Available at https://www.radicalphilosophy.com/article/axiomatic-heresy (accessed November 2022).

Brassier, R. 2007. *Nihil Unbound*. Basingstoke and New York: Palgrave Macmillan.

Brassier, R. 2009. *Against an Aesthetics of Noise*. Available at https://www.ny-web.be/artikels/against-aesthetics-noise/ (accessed August 2022).

Bratton, B. 2020. 'THE PLAN', in Garayeva-Maleki, S. and Munder, H. (eds), *Potential Worlds: Planetary Memories & Eco-Fictions*. Zurich: Migros Museum für Gegenwartskunst, YARAT Contemporary Art Space and Scheidegger & Speiss.

Bryant, L., Srnicek, N. and Harman, G. (eds). 2011. *The Speculative Turn: Continental Materialism and Realism*. Melbourne: re.press.

Capra, F. 1996. *The Web of Life: A New Scientific Understanding of Living Systems*. New York: Anchor Books, Doubleday.

Capra, F. 2014. *The Systems View of Life: A Unifying Vision*. Cambridge: Cambridge University Press.

Coole, D. and Frost, S. 2010. *New Materialisms: Ontology, Agency, and Politics*. Durham and London: Duke University Press.

Cox, C., Jaskey, J. and Malik, S. (eds.). 2015. *Realism Materialism Art*. London: Sternberg Press.

Deleuze, G. and Guattari, F. 1986. *Kafka: Toward a Minor Literature*. Minneapolis, MN: University of Minnesota Press.

Dolphijn, R. and van der Tuin, I. 2012. *New Materialism: Interviews & Cartographies*. London: Open Humanities Press. Available at http://openhumanitiespress.org/books/download/Dolphijn-van-der-Tuin_2013_New-Materialism.pdf (accessed July 2022).

Galloway, A.R. 2006. *Protocol: How Control Exists after Decentralization*. Cambridge, MA: MIT (Leonardo).

Galloway, A.R. 2010. *French Theory Today: An Introduction to Possible Futures. A pamphlet series documenting the weeklong seminar by Alexander R. Galloway at the Public School New York in 2010. Pamphlet 4 | Quentin Meillassoux, or The Great Outdoors*. Available at http://cultureandcommunication.org/galloway/FTT/French-Theory-Today.pdf (accessed July 2022).

Galloway, A.R. 2014. *Laruelle: Against the Digital*. Minneapolis, MN: University of Minnesota Press.

Haraway, D.J. 2016. *Staying with the Trouble: Making Kin in the Chthulucene*. Durham and London: Duke University Press.

James, R. 2019. *The Sonic Episteme: Acoustic Resonance, Neoliberalism and Biopolitics*. Durham and London: Duke University Press.

Kho, Z.Y. and Lal, S.K. 2018. 'The human gut microbiome – A potential controller of wellness and disease'. *Frontiers in Microbiology*, 9(1835). doi:10.3389/fmicb.2018.01835

Laruelle, F. 2010a. *Future Christ: A Lesson in Heresy*. London and New York: Continuum.

Laruelle, F. 2010b. *Philosophies of Difference.* London and New York: Continuum.
Laruelle, F. 2010c. *Philosophie Non-Standard: Generique, Quantique, Philo-Fiction.* Paris: KIME.
Laruelle, F. 2012a. *The Concept of Non-Photography.* Falmouth and New York: Urbanomic.
Laruelle, F. 2012b. *Photo-Fiction, A Non-Standard Aesthetics.* Minneapolis, MN: Univocal Publishing.
Laruelle, F. 2012c. *From Decision to Heresy.* Cambridge, MA: MIT Press.
Laruelle, F. 2013a. *Anti-Badiou: On the Introduction of Maoism into Philosophy.* London and New York: Bloomsbury.
Laruelle, F. 2013b. *Philosophy and Non-Philosophy.* Minneapolis, MN: Univocal.
Laruelle, F. 2013c. *Principles of Non-Philosophy.* London and New York: Bloomsbury.
Lovett, M. 2021. 'Towards a quantum theory of musical creativity', in Hepworth-Sawyer, R., Hodgson, J., Paterson, J. and Toulson, R. (eds), *Innovation in Music: Future Opportunities.* London and New York: Routledge.
Malm, A. 2018. *The Progress of This Storm: Nature and Society in a Warming World.* London and New York: Verso.
Margulis, L. and Lovelock, J. 1974. 'Atmospheric homeostasis by and for the biosphere: The Gaia hypothesis'. *Tellus A* 26, pp. 2–10. doi:10.1111/J.2153–3490.1974.TB01946.X
Marx, K. [1859]1904. *A Contribution to the Critique of Political Economy.* Chicago, IL: Charles H. Kerr.
Meillassoux, Q. 2008a. *After Finitude.* London and New York: Continuum.
Meillassoux, Q. 2008b. 'Time without becoming'. Available at https://speculativeheresy.files.wordpress.com/2008/07/3729-time_without_becoming.pdf (accessed July 2022).
Morton, T. 2013. *Hyperobjects: Philosophy and Ecology after the End of the World.* Minnesota, Minneapolis: University of Minnesota Press.
Morton, T. 2018. *Being Ecological.* London: Penguin Random House.
Morton, T. and Boyer, D. 2021. *Hyposubjects: On Becoming Human.* London: Open Humanities Press.
Naess, A. 2008. *Ecology of Wisdom.* London: Penguin Random House.
Negarestani, R. 2014. 'The labor of the inhuman', in Mackay, R. and Avanessian, A. (eds), *Accelerate: The Accelerationist Reader.* Falmouth: Urbanomic Media.
Negarestani, R. 2018. *Intelligence and Spirit.* Falmouth: Urbanomic Media.
Negarestani, R. 2020. 'On an impending eternal turmoil in human thought,' in Garayeva-Maleki, S. and Munder, H. (eds), *Potential Worlds: Planetary Memories & Eco-Fictions.* Zurich: Migros Museum für Gegenwartskunst, YARAT Contemporary Art Space and Scheidegger & Speiss.
Rekret, P. 2016. 'A critique of new materialism: Ethics and ontology'. *Subjectivity*, 9, pp. 225–245. doi:10.1057/s41286-016-0001-y
Rekret, P. 2018. 'The head, the hand, and matter: New materialism and the politics of knowledge'. *Theory, Culture & Society*, 35(7–8), pp. 49–72.
Rovelli, C. 2021. *Helgoland: The Strange and Beautiful Story of Quantum Physics.* London: Allen Lane.
Smith, A.P. 2013. *A Non-Philosophical Theory of Nature: Ecologies of Thought.* New York: Palgrave Macmillan.
Toscano, A. and Kinkle, J. 2014. *Cartographies of the Absolute.* New York: Zero Books.

2

MUSIC AND MATERIAL CREATIVITY

In *All Art Is Ecological*, Timothy Morton describes ecological art as 'art that includes its environment(s) in its very form' (Morton, 2018: 18). Given that Morton sees practices of art production as being entangled with the myriad contexts and ecosystems in which artmakers are themselves enmeshed – for example race, class and gender (ibid.) – we can certainly make meaningful connections between Morton's work and the new materialist milieu that we encountered in the last chapter. Morton's onto-epistemological model – a framework for understanding what the world is (ontology), and for understanding how our own understanding of the world operates (epistemology) – sees the world as an actually-existing-world that we have a real, concrete relationship with. Unlike Meillassoux, who proposes that we are able to conceive of an absolute real that is not present in the world, but nevertheless can be conceptualised from our position in the world, Morton's is a strain of speculative realism that articulates the reality of things, rather than simply the reality of an absolute that we are able to conceive of from our position in the world of things. Morton and Meillassoux are alike in their reasoning that absolute reality is not simply available and accessible to us, which is to say that we can't simply reach out and touch or hear the reality of things in the world around us. The key difference is a question of positioning. It is a question of whether the real is in the world that we are part of, or if it is outside the world, but that somehow, we are able to acknowledge its existence in a non-problematic way. The latter is Meillassoux's view, whereas for Morton, given that art – because of its ecological nature – is part of the actually-existing real world, art is also therefore something that is real, and it follows that both art and the world really are a part of each other. Morton goes on to reference correlationism, and articulates a progression from postmodern, correlationist thought, to the ecological perspective that they are arguing for. Morton's version is more nuanced

DOI: 10.4324/9781003225836-3

than Meillassoux's, and they foreground a subtle, but vital distinction between theirs and Meillassoux's approaches: 'correlationism is true: you can't grasp things in themselves' (ibid.). Thus, Morton asserts that although we may not be able to grasp things in themselves, this does not mean that things do not exist separately from humans' capacity to grasp them.

Current philosophical theory that operates in this post-correlational mode, variously labelled as new materialism, speculative realism, object-oriented ontology or non-philosophy, includes a range of strategies that are being used to move away from what Morton sees as a post-Kantian tradition of correlationist thought. As we have seen, Morton describes 'subscendence' and 'implosive holism' as terms designed to convey how wholes (for example, humans) are smaller than the sum of their parts. Morton is pushing back against the common idea that wholes are greater than the sum of their parts, which they see as a corruption of the formulation in gestalt psychology where wholes are different from the sum of their parts (Morton, 2018: 101). For example, the term 'humankind' cannot encapsulate all of the things that it brings together into a supposedly complete whole, particularly when our own microbiomes partly consist of non-human biota. Morton also makes reference to how we are able to modify ourselves with electronic prostheses (Morton, 2017), leading them to suggest that the category 'human' is not a whole that contains and gives meaning to all of its constituent parts. They therefore see objects as partial objects, and an 'implosive whole' is an object that is always less than its parts. As a result, although we humans may well appear to be individual, self-contained things, we are in fact always symbiotic and each of us is a member of many communities. As ecological beings, all of us have what Morton calls 'spectral halos' (ibid.): the myriad parts that make up and compose what appears to be a self-same whole.

This chapter draws on this kind of ecological thinking, which has arisen in the wake of the various perspectives explored in the last chapter, and brings this thought to bear on the production and experience of music. In this sense, we are now fully embarked on the task of mattering music. Before continuing however, as a caveat to what follows, it is important to acknowledge that the areas of thought we are drawing on are also directed towards demonstrating that theoretical framing and philosophical analysis cannot explain creative practices such as music. Neither do these theoretical positions suggest that it is possible to ventriloquise music, by putting into words what creative practices such as music are unable to communicate themselves. In *More Brilliant than the Sun*, Kodwo Eshun wrote:

'Instead of theory saving music from itself, from its worst, which is to say its best excesses, music is heard as the pop analysis it already is [...] Far from needing theory's help, music today is already more conceptual than at any point this century' (Eshun, 1998: 00-004-003).

Similarly, for Gilbert and Pearson, in *Discographies: Dance, Music, Culture and the Politics of Sound*:

> Music is constituted by waves of sound that vibrate through the air, vibrating the eardrum in specific patterns which are registered by the brain [...] These vibrations are registered on some level throughout the body [...] When a 25kW bass-line pumps through the floor and up your legs, you know that music isn't only registered in the brain [...] all sound is registered on a fundamentally different level to language or modes of visual communication, just as the philosophical tradition has always suspected (and feared).
>
> *(Gilbert and Pearson, 1999: 44)*

Increasingly, such perspectives have worked to reconfigure the relationship between the material experience of music and art as somatic practices, and philosophy, emphasising how music and art generate concepts, affective experiences and ways of thinking that do not need to be – and in fact cannot be – explained by philosophy. Indeed, in his reading of François Laruelle's conception of non-standard aesthetics, Thomas Sutherland suggests that

> a truly immanent art [is] art as an immanent act, an act that is immanent to itself, neither representative nor expressive. A non-conceptual form of art, that does not need to extract concepts from elsewhere, nor to have such concepts impressed upon it, is an art that is already aestheticized, already given.
>
> *(Sutherland, 2016: 65)*

This is to say that, for Laruelle, art is wholly self-contained. Art does not produce meaning because it is referring to something else. Neither does it need to be interpreted in a certain way for its meanings to be apprehended. For Sutherland, Laruelle is creating an 'artistic thought; a thought according to art, rather than a mere philosophy of aesthetics [...] it is the question of whether art can produce an art of philosophy, rather than a philosophy of art' (Sutherland, 2016: 65). Essentially, Laruelle is asserting a reciprocal, non-hierarchical relation between art and philosophy, a key principle of his non-philosophical objective to demonstrate that philosophy has no more access to, privileged understanding of, or capacity to produce the Real, than any other form of human endeavour. Philosophy can no more explain the real nature of the world than can art or music, it is simply a practice and must take its place alongside music, art, cooking, motor racing and any number of human practices that we use to navigate and to create experiences in the world.

Although there are clear lines of difference between their approaches, the thinkers that we have engaged with position the relation between an absolute reality and the concrete experiences of the everyday world in such a way that belies a certain kinship, in terms of the material presence of the absolute real. For Laruelle, the Real is always immanent to, and therefore never 'not present', in the real world. For Meillassoux, the outside of thought is something that we are aware of, even if it is non-conceptualisable. For Morton, everything is real, and the singularity of objects is always present, even if we are not able to access their realness. For Barad – while they share Laruelle's and Morton's principle that at no point is anything *not* real – real experience is produced by the constant interactions between everything in the world; measurements, entanglements, diffractions and the marking of bodies. The following engagements with music acknowledge these approaches, in the way that listening to and thinking about music are framed as sites of discursive and material innovation which do not need to be explained by philosophical positioning. Instead, the trajectory of this chapter is to approach music with an awareness of theories of ecology and interdependence in order to articulate what music is already doing. While music continuously connects with other discourses and practices in order to evolve and create new patterns in experience and thought, it does not need these others in order to validate its capacities. On one hand, the somatic character of music as sound requires us to think differently about how we experience music as a material-discursive presence in our bodies. On the other, the recalibration of critical and philosophical theorisation in relation to music's capacity to generate concepts and modes of thought, which increasingly suggest that music is immanent to itself – 'neither representative nor expressive' (ibid.: 65) – demands that we reconsider our onto-epistemological habits. Just as Naess indicated via the ethics of deep ecology, the challenge is to reconsider what we think things are – both ourselves and the world around us – and carefully reformulate how we understand our relationship to the world we inhabit. Such challenges are evidence of a growing understanding of music's capacity to create knowledge and generate experience free from the need of being explained by anything else, and free from needing to be fed ideas by philosophy in order to create new forms and new modes of thought.

In the Introduction, I discussed musics that have been made within the context of ecological perspectives; the music of John Luther Adams is actively directed towards exploring natural phenomena, and in a broad sense operates as an enquiry-in-sound into human relationships with non-human natural environments. Similarly, we encountered, in acoustic ecology, recording and listening practices that have been developed as a means to identify and document the heterogeneous and fragile nature of the natural world, along with the sonic incursions of our industrial society. We also acknowledged the work of ecomusicologists, including Kyle Devine, Matt Brennan and Mark Pedelty, who have helped to shine a light on the ecological nature of music as a material practice, and its environmental impacts. In this chapter, the emphasis shifts, as the music we hear foregrounds its ecologicality in different ways. Importantly, as we focus in on practices of making and listening to music, we hear that music does not need philosophical theory to tell us about how ecological and interdependent it is, nor does it need to be music that is about ecology or ecological issues in order for it to be ecological. As we listen to the work of Jeremiah Chiu and Marta Sofia Honer, to Ralph Vaughan Williams, to Bendik Giske, John Butcher, Rhodri Davies, Mark Guiliana and Makaya McCraven, what becomes increasingly apparent is that, to borrow Donna Haraway's term, thinking about composing as a process of *composting* is simply a way to recognise what the process of making music already is. Similarly, when we think from a listener's perspective, we realise that listening – as a creative act – is an ecological practice; to listen is to be enmeshed in sounds and meanings, but it is also a point of meaning making, an agential cutting and splicing procedure that produces real, non-representational meaning in the world. As such, my interest here is not to explain music in terms of new materialism and speculative realism; instead, the trajectory of the chapter is to *matter music*, to think about and listen to what can come when we tune ourselves to ecological and interdependent practices of patterning sound.

The chapter ends by engaging with two theories of practice; the composer Pauline Oliveros' concept of quantum listening, an extension of her practice of deep listening, and Laruelle's non-philosophical concept of fiction, an approach to thinking of creative practice as a means to create something that is not wholly of our own making. Just as Meillassoux suggests that we are able to imagine an absolute reality that exists beyond to our capacity to experience it, Laruelle indicates that through creative practices – such as music making – we can not only imagine such a world, we can produce it. Again, following the trajectory of this book to engage directly with the interdependent materiality of creative music practices, these theories are not intended to *explain* music in any way. Indeed, as Laruelle repeatedly asserts, it is the capacity of creative practice to continually exceed philosophy's attempts to explain it – or maintain a hierarchical relationship with knowledge of the world – that demonstrates music's interdependent relationship *with* theory, rather than its dependence *on* theory.

Composing as Composting

Released in 2022, Jeremiah Chiu and Marta Sofia Honer's album *Recordings from the Åland Islands* (Chiu and Honer, 2022) is a suite of pieces that were assembled from recordings the pair made during visits to the Åland Islands, an autonomous territory of Finland, comprising over 6,000 islands in the Baltic Sea. The source material for the album is field recordings, along with recordings of improvisations the pair made whilst staying at the Kumlinge municipality, including performances in Kumlinge's fourteenth-century church. The music on *...Åland Islands* is made out of the various sonic environments that Chiu and Honer explored and created during their stays at Kumlinge, and the transparency of their construction lays open their ecologicality: this is music that is ecological in the fabric of its making. Furthermore, because *...Åland Islands* focuses our attention on the soundscapes of Kumlinge, as with Truax's and Schaefer's work, it is also *ecomusicological* in an acoustic ecological sense. In his review of the track 'Snåcko', Philip Sherburne writes that 'the music feels less like a composition than a snapshot of the landscape' (Sherburne, 2022). The *ecomusicologicality* of 'Snåcko' springs from this meshing of field recordings and musical performances, of the aestheticisation of nonmusical natural and human-made sounds, along with Chiu and Honer's creative process, which creates an integrated sonic whole from these diverse elements. It is also ecomusicological in the way that it draws on histories of field recording, of improvisation, of site-specific performance in order to be something that is simultaneously all of those things, and yet is also itself; it is music that is ecological in its making, and is also about being ecological. However, perhaps the most important thing to recognise in Chiu and Honer's music is that it is ecological and ecomusicological, not because philosophy tells us that it is so. Its ecologicality stems from the material processes of its making. Sound recording technology will always record the sounds of any environment it is being used in, and musicians can always make music wherever there are instruments. In the album, Chiu and Honer simply foreground these processes and, in so doing, they aestheticise the making processes and the locatedness of the album's production. These processes are happening as part of all recorded music, the difference is that many producers choose to conceal the locatedness of production, creating music as fictitious sonic worlds. However, as we shall discuss later in the chapter, music fictions are no less true, and no less materially grounded in processes of making than *Recordings from the Åland Islands*.

Although 'Snåcko' makes its construction more evident than other musics might do, it is no more ecological than others; its transparency simply makes its interdependent nature more apparent. It is interdependent because it is made from a broad range of histories, ideas and sounds, and at the same time its existence feeds back into those component parts, adding to them with the new contexts and perspectives that result from being part of the *Recordings from the Åland Islands* project. For example, while the music on *...Åland Islands* may indeed ask questions about what it is that we are listening to – Music? Improvisation? Field recording? Some kind of hybrid? – it also asks a broader question about what composition in itself is. What kind of process is the production of music? Is it simply about assembling different sounds together, or do different kinds of sounds direct us to listen to them in certain ways? Can every sound be made into music, simply by being placed within the frame of a composition? *Recordings from the Åland Islands* lays such questions open by dint of its very obvious assemblage of heterogeneous – even multispecies – sonic components. There are birds on the album, there are snatches of conversation and sounds of the sea. There are impromptu performances on a church piano, improvisations on a viola, electronic sounds, along with the studio-based edits and compositional choices about which sounds to use, and how to make them work together as a track. We listen to the tracks on *...Åland Islands* as composites of all of these things and more, and yet Chiu and Honer's performances, recordings and compositions are not simply references to, or representations of, these myriad components. The tracks ask us to think what they are, about what it is that we are listening to, but they also ask questions about how to listen. What are our expectations of the listening process? Listening, as much as making, is therefore part of the ecology of *Recordings from the Åland Islands*. Nothing is fixed: many listenings are possible, all based on concrete relations between material listeners and material music.

On the sleeve notes of his 1986 album *On Land*, Brian Eno describes aspects of the record's production, a process not unlike the one used by Chiu and Honer. Eno relates how his source material was drawn from field recordings (including rooks, frogs and insects), electromechanical and acoustic instruments and non-instruments (including lengths of chain, sticks and stones), along with his own catalogue of releases. His sense was that some of his earlier pieces had been 'digested' by the more recent ones. He described this production technique as 'composting; converting what would otherwise have been waste into nourishment' (Eno, 1986). As already mentioned, the multispecies feminist theorist Donna Haraway also alights on the process of composting:

> We are compost, not posthuman; we inhabit the humusities, not the humanities [...] Critters – human and not – become-with each other, compose and decompose each other, in every scale and register of time and stuff in sympoietic tangling, in ecological evolutionary developmental earthly worlding and unworlding.
>
> *(Haraway, 2016: 97)*

Composing as composting, sounds and music being digested (and regurgitated?), new music always being made of old music, and of everything else that its maker has come into contact with. Music literally feeds on the things that it is made from. This is a compelling image with which to think about not only Chiu and Honer's creative process, but creative processes in a wider sense. Beyond thinking about composting as a fundamentally co-creational, co-evolutional process, Haraway engages with creative production as processes of pattern-making. She describes how Navajo string games, which consist of weaving lengths of string into different patterns that are passed back and forth between players' hands (a version of which exists as the well-known string game 'cat's cradle'), are processes of thinking as well as making. Haraway builds on this image, bringing together 'science fiction, speculative feminism, science fantasy, speculative fabulation, science fact, and string figures' (Haraway, 2016: 10) – all variations of 'SF' – to suggest that 'playing games of string figures is about giving and receiving patterns' (ibid.: 14). In this light, creative acts are processes of thinking and making new patterns, of making new ways to share patterns with others. Haraway's idea here is that the simple process of creating new patterns from a short length of looped string, entwined across the digits of two sets of human hands, is an image of the essence of creative practice: giving form to basic materials, problem solving (how to form what is given into a new, meaningful shape) and a passing of the creative work to another human (even if this is an imagined audience, part of the creative process is about creating some kind of externality that helps to shape and complete the work, even if this is other creative works). Thus, making and thinking are bounded, located practices that make with the components of the world; string and hands, recordings of birds and viola improvisations. Where Haraway enables us to think in a more focused way about creativity as composting, pattern making and worlding, a process of creating the world, by creating worlds within the world, the new materialist and speculative realist perspectives build coherent frameworks that emphasise the real nature of these worlds.

Eno's *On Land* project was one of sonic worlding; it was designed to evoke and capture, through sound, a sense of place, without evoking or referring to a particular place; of locatedness, without being located. Indeed, as Eno informs us, he felt 'no sense of obligation to realness' (Eno, 1986). We might like to think of Eno's sonic worlding as 'speculative fabulation', for although he did not feel obliged to align his approach with any sense of representationalism (what he calls 'realism'), this by no means detracts from the actual realness of Eno's music. The music on *On Land* was – and remains – something entirely real and material, something that – in itself – creates real effects in the world. It is the subject of multitudes of real experiences, and it is part of an ongoing topology that we experience as the real world. The point is, that *On Land* is not a simple representation of a particular place or sonic environment. On *Recordings from the Åland Islands*, we very much encounter such a sense of non-representational realism. For example, on 'Stureby House Piano' we hear a piano with an exchange of muffled human voices in the background. As the track progresses, the voices disappear, and the piano sound is processed with delay and what sounds like time stretching, and overlaid with synthesizer sounds, foregrounding the constructed nature of the music we're listening to. At 2′30″, the piano begins to repeat a simple, disjointed arpeggio pattern, which is joined by a bell-like synthesizer pattern. There is certainly an aspect of this section of the track that bears relation to change ringing patterns used by bell-ringers, where a small number of tones are repeated in different sequence over and over again. There is a further shift in the track's arrangement, that occurs between approximately 3′50″ and 4′05″, where different elements of recorded piano are overlaid on top of each other. The final section of the track consists of what sounds like another extract of improvised piano, comprising three note arpeggios and short melodic and chord fragments, and ending quietly with a rising minor 7th arpeggio that makes subtle reference to Debussy's 'Girl with the Flaxen Hair'. The track makes a final transition to another fragment of conversation, where we can just discern one voice saying, 'it sounds nice, but you should have recorded this…', to which a second replies, 'I just did buddy!' (Chiu and Honer, 2022). While we cannot be sure what aspects of the recording are true documentations of performances that occurred at fixed times and places, as with the real musical occurrences that were the product of Eno's rejection of realness, Chiu and Honer's music brings into sharp focus the concrete process of music making. In listening to 'Stureby House Piano', we ourselves are listening to Chiu listening; listening to him making choices about what fragments of his recordings will be used, and where they will be placed. Indeed, we may well be listening to him talking to his project collaborators about the process of making the recordings that populate *Recordings from the Åland Islands*. We are listening to the process of making music as a process of listening and talking about how to make music. In this sense, even listening becomes entangled in the process of performance and composition as composting.

While the highly transparent nature of *...Åland Islands* makes it rich in opportunity to examine the ecological process of making and listening to music, we can see and hear such process at work in music from earlier epochs. 'The Captain's Apprentice' is a traditional song that Ralph Vaughan Williams collected in Kings Lynn, Norfolk in 1905. The words of the song relate the story of a ship's captain taking on a young apprentice to save him from the hardships of St James' workhouse. One day, the young man offends the captain, who proceeds to beat and torture the boy, leading to his death. Vaughan Williams' setting of the song was published in 1908, part of his first published collection, *Folk Songs from the Eastern Counties*. In his liner notes to the 1976 recordings, Michael Kennedy describes Vaughan Williams as a 'melodic archaeologist', explaining how the latter's song collecting activities in the early twentieth century were driven by a desire to capture, preserve and 'give back to the people' a set of musical traditions that were held in the memories an older generation of labourers and sailors, before they were lost to posterity (Kennedy, 1978, in Vaughan Williams, 2008). Written in 1906, and revised in 1914, Vaughan Williams' 'Norfolk Rhapsody No.1' arranges the melody of 'The Captain's Apprentice' for instrumental forces. The piece begins in E minor, with the upper strings voicing a fragment of the song's melody in a repeated D, A, B phrase, before the viola enters after a bar's silence approximately one minute into the piece. After its haunting entry, the song's melody is then orchestrated across the strings and woodwind from 4′30″ (Vaughan Williams, 1991). We can draw parallels between Vaughan Williams' song-collecting practices and more contemporary sound-recording processes, used to populate many song archives. Similarly, his instrumental arrangement of 'The Captain's Apprentice' can be likened to digital sampling techniques. With deference to Eno and Haraway, in ecological terms, we can also understand Vaughan Williams' practices as pre-digital musical composting. Swathes of musical histories have been collected and refashioned by the likes of Vaughan Williams, Cecil Sharpe, Bela Bartok, Pete Seeger, Marley Marl, DJ Shadow and many more, and while it is often tempting to hear the music that results from collecting as the creative act, collecting is also a far from neutral process. Transcribing or recording traditional songs, listening to music and selecting key passages from which to create a sample, archiving and cataloguing music are all part of the composting process.

Indeed, the vitrification and storage of traditional dances and songs in archives such as The English Folk Dance and Song Society at Cecil Sharp House in London (EFDSS, 2022) or the *Folk Songs* collection at St Fagans National Museum of History in Cardiff (Amgueddfa Cymru, 2022), are processes full of creative decision-making – what to preserve, how to preserve it, how to make use of archival material – and, what is more, these stores of cultural memory are still being used to propagate the production of new culture.[1] Archives as compost heaps is an attractive image; centuries of culture being both stored and converted into nourishment for future generations, to not simply rediscover, but to use, and be shaped into new patterns which themselves pattern their own contemporary discourses.

The drummer and producer Makaya McCraven has developed a creative practice that also involves recording improvised performances, and creating new pieces by editing, looping and layering these recordings in different ways. McCraven will record group performances, in a variety of settings, ranging from concert venues to musicians' houses, forensically combing through the recordings to identify certain moments and passages that he then re-assembles into new, stand-alone pieces. Working with the record label International Anthem, McCraven hosted and recorded a series of improvised jazz shows during 2012–13, which resulted in 48 hours of recordings (McCraven, 2022). He then mined these recordings, extracting phrases and beats, along with longer passages, before reassembling them into new pieces for the album *In The Moment* (2015). Across a subsequent series of albums, including *Where We Come From* (2018), *Universal Beings* (2018), *Universal Beings E&F Sides* (2020) and *In These Times* (2022), McCraven went on to hone his concept and approach, working in small group settings in London, New York, Chicago and Los Angeles. This international, highly collaborative process reflects his view that 'we are all interconnected, we're all building off the past. It's a vision of a world that's less fragmented and segregated [...] What is a place? Other than the people. It's just dirt, you know?' (McCraven, in Joyce, 2018). McCraven's sense of interconnection and shared histories, along with a more specific focus on musical cultures, is also redolent of Haraway's and Eno's ideas. As compost, McCraven's music is completely infused with histories; personal and social human trajectories, as well as musical, specifically jazz, legacies; although he does not view his music as an homage to jazz as such. Instead, he suggests that 'what I'm doing is [not] necessarily that far off of the legacy of jazz that I grew up in' (McCraven, in Kerson, 2019), an indication that his music is an extension – a further composting – of jazz's creative principles, rather than a celebration of them.

On closer analysis, McCraven's music also opens a wider range of ideas about music and practices of music making. In essence, the tracks on the four albums listed above are created by sampling recordings of musicians' performances that happened at certain times and in certain places. Although McCraven professes not to be concerned with place in the sense of cities, or even continents, what he has recorded are performances in environments where there are people; both musicians and audiences. On tracks such as 'Run 'Dem' and 'Too Shy' from *Where We Come From*, the looped patterns – of drums, keyboards, tuba or otherwise – frequently contain the audience responses to a given musical moment. Thus, during the opening minute of 'Run 'Dem', a shout of 'woo!' and a high-pitched celebratory whistle are repeated as the track builds towards a transition point into a new section at 1′08″ (McCraven, 2018b). Similarly, on 'Too Shy', another whistle from the audience makes up part of the looped components of the track. These tracks imply that – unlike his forebears in hip hop, drum'n'bass and other electronic dance musics, who often worked to create a drum loop out of an isolated fragment of a track that was free from any other sounds or instruments – McCraven is interested in sampling whole moments of time, and using all of the sonic elements of a given moment in the creation of new music. Not only does this approach connect with Eno's ideas about the place of realness in the construction of environments (both spatial and temporal), but it is also relevant to how time is both captured and produced by things, sometimes communities of things; a theme we shall turn to later in the chapter. 'Black Atlantic' is another audible example of sonic composting. In this instance, there are no audience sounds; the track is made purely from fragments of instrumental performances. A repeated drum and keyboard phrase serves as the musical foundation for the piece, while Shabaka Hutchings' tenor saxophone, Tomeka Reid's cello and Junius Paul's double bass are all looped in various ways, including short rhythmic fragments that last less than one second, and repeated arco phrases that are digitally pitched up and down to give the track a sense of forward motion. The cohesiveness of McCraven's soundworld across these albums suggests that many, if not all, of his tracks are made in this way; recording performances in particular spaces, at particular times, and with certain other people in the room (musicians and/or audiences), and then fragmenting and reforming these spatiotemporal sounds into new music. As with the discussions of Vaughan Williams' and Chiu and Honer's music, McCraven's collecting/recording/sampling/composing-as-composting process makes explicit the discursive-material nature of creative sound and music practices; these are ecologies of making and listening, of histories and futures, of people, places and music.

Another variation on this theme is the album *Telyn Rawn* by the harpist Rhodri Davies. Between 2016 and 2018, Davies worked on a project to recreate a 'telyn rawn', a small lap harp strung with horse hair that would have been in use in medieval Wales (Clera, 2021). The telyn rawn was gradually supplanted by the 'telyn deires' (the triple harp) during the seventeenth century, and one of the last recorded references to the horse hair harp was in 1802 (Davies, 2020). Released in 2020, the album *Telyn Rawn* is a set of 18 pieces that Davies 'improvised without any melodic, rhythmic or musical material [being] decided upon prior to the moment [they were] played live in the studio' (ibid.). However, Davies had researched the history of the telyn rawn, learning about its construction and tuning conventions, and worked through centuries-old musical scores, such as the

Robert ap Huw manuscript, which is thought to have been played on a telyn rawn (ibid.). He lists 'designing and building a long-forgotten instrument, experimenting with wound and pleated horse hair strings, engaging with historical texts and poetry, learning the techniques from the Robert ap Huw manuscript, researching the importance of the horse and horse cults in Welsh culture and attending local folk sessions at Ty Tawe in Swansea' as part of his preparations for recording the album (ibid.). As a musician who is closely associated with contemporary classical and improvised musics, along with alternative folk and pop musics (Davies, 2022), Davies is well placed to locate the telyn rawn within an intersecting set of musical and historical trajectories:

> The further back in time we look, we are left with a fragmented picture because of the scant documentation that survives [...] The vernacular and improvised music of the day disappeared because it was not sacred or court music. Improvised, everyday, unusual or unorthodox music easily vanishes: and of course, that is the nature of improvisation.
>
> *(Davies, 2020)*

Out of these entangled histories, practices and ideas, a picture emerges of a musician who has immersed himself in the technologies and techniques of a long-forgotten instrument, contextualising his findings in terms of wider cultural symbols and practices, while at the same time drawing a line that connects contemporary improvised music making to its own history of creative musical practice. Throughout his career, Rhodri Davies has made extensive explorations of the affordances of the harp as a musical instrument, playing different kinds of harps, playing modified harps, playing harps with a variety of techniques and preparations, as well as burning, freezing and burying harps in various rituals, performances and installations (Davies, 2022). The pieces on *Telyn Rawn* therefore also draw on his own knowledge and experience of pushing the harp to the extremes of what it can do as a both a musical and cultural object. As a suite, these pieces are Davies' musical explorations of the affordances of the telyn rawn as an instrument, and a meshing of the telyn rawn with his own knowledge of harp-playing across a range of styles, technical approaches, cultural and historical variations. The album's opening track, 'Penrhiw', has a symmetrical structure, beginning and ending with a pulsing two-note plucked chord pattern, which Davies varies by changing the top note. The middle section features short melodic phrases, that grow out of arpeggios and chord groupings, and the piece ends with a silent pause before the pulsed chords return, leading us to a short final arpeggio gesture. 'Angharat Ton Uelen' makes use of a number of rhythmic elements, including a short section in which Davies makes use of very fast picking, giving the impression that the piece is shaped by the physical relationship between his fingers and the harp strings, and even by a set of movements and patterns that the fingers fall into, as much as it is a melodic and rhythmic construction. A number of tracks on *Telyn*

Rawn are played using a horse hair bow, including 'Afon Dewi Fawr', 'Waunceiliogau' and 'Penglog Ceffyl'. They each explore the harp's capacity to create groups of tones and sounds alongside or on top of a repeatedly bowed drone. In 'Afon Dewi Fawr' we hear subtle overtones and harmonics, possibly as Davies positions and moves his fingers along the string as it is being bowed. 'Waunceiliogau' comprises a repeatedly bowed chord. The bowing on this piece generates a number of sounds in addition to the repeated chord tones, including the abrasive rubbing sound of the two sets of horse hairs (bow and strings) moving against each other and the occasional squeak that this motion creates, along with quiet harmonics that arise from the vibrations of the strings. The bowing on 'Penglog Ceffyl' – which translates as 'horse skull' – implies a much faster tempo, and double-speed movement against the harp's strings. Again, Davies' technique brings out harmonic overtones over the single bowed fundamental note. Amongst the other plucked pieces, 'Errdigan Gwawr' is the most overtly 'folk' sounding piece on the album, and features a skipping 6/8 rhythm, that oscillates between a G major chord and C major chord, where the bass note remains on a constant G throughout the piece, and the higher notes move around to adjust the harmony. As Davies comments, 'the sound of the telyn rawn is different to the contemporary harp but the timbre is not foreign to musicians working with free improvisation' (ibid.), which reinforces the sense in which the *Telyn Rawn* project is a meeting of material forces across time and in space: centuries-old techniques and traditions being encountered by Davies' contemporary musical sensibilities, and an ancient instrument with all the technical and practical constraints accorded by its relatively primitive design being played by someone with a rich and wide-ranging knowledge of musical techniques and styles. *Telyn Rawn* in no way sounds like a representation or recreation of ancient harp styles, nor is it simply a modernist's attempt to bring a medieval instrument up to date. It is a composting of physical materials, histories, techniques, knowledges and creative impulses.

In their response to Olafur Eliasson's *Ice Watch* – a piece of public art installed at the United Nations Conference on Climate Change, which consisted of 12 blocks of ice, each weighing over 6 tonnes and arranged so as to represent the 12 hours on a watch face (Ice Watch, 2015) – Timothy Morton offers a compelling perspective on the entangled, non-human nature of time. Proposing that 'everything emits time, not just humans', Morton draws attention to how humanity has imposed a single, human-level timescale on a planet that consists of a multitude of timescales all constantly interacting with, and diffracting each other (Morton, 2018: 64), suggesting that hubris has led to the human-altered climate of the Anthropocene, as highlighted in Eliasson's work. While Morton recognises that *Ice Watch* was a visual joke – a watch made of ice, and we're watching the ice melt as we run out of time because of climate change – they also see it as a demonstration of how time is not only something that humans make. The world is made of things that do not simply operate according to a human timescale, and they point out that the sun, the Paris climate and the

Paris drainage system are all interacting with the 12 ice blocks on timescales that are distinct from human time. These non-human interactions are therefore stark reminders that things produce time itself; our human timescales are not the measure of all things. So saying, Morton's analysis therefore reminds us that music – just like ice – does not just exist in time; it produces time. As a time-based phenomenon, music not only has a duration, which can be measured against already existing clock-time, it also creates its own timeframe: it times its listener. Just as music makes marks on bodies – the experiences it engenders when we listen to it, or even the damage it causes to our ear drums when we listen to music at excessive volumes – music also creates time. There is a resonance here with one of Karen Barad's key ideas, their conception of how processes of boundary-making are another way to understand processes of meaning making: 'apparatuses produce differences that matter – they are boundary-making practices that are formative of matter and meaning, productive of, and part of, the phenomena produced' (Barad, 2007: 146). What is both important and compelling about Barad's approach in their conceptualisation of boundaries and boundary-making, is that boundaries do not pre-exist the interaction of bodies. Boundaries are produced through the process of inter-/intra-action. The importance of this cannot be over-emphasised in Barad's work, since it is the proposition that supports their claim for the existence of non-subjective truths, for an actually-existing real. In this regard, boundaries – like the marks on bodies that are the traces of entanglement – are another way of saying 'spacetimemattering'.

A musical instrument is an apparatus, as is a music production tool; a studio, a computer or a microphone. Jeremiah Chiu's recording equipment is an apparatus, so too is Rhodri Davies' telyn rawn and Vaughan Williams' manuscript paper which he used to transcribe 'The Captain's Apprentice' in 1905. Later in the chapter, we shall listen to the music of Bendik Giske, whose saxophone and the directional microphone array that he uses is also an apparatus, along with the Vigeland mausoleum in Oslo, a reverberant chamber in which Giske performed the pieces that became his first album, *Surrender*.[2] As such, Barad's images of 'marks on bodies' and 'boundaries' are indications that making and playing violas, harps, saxophones and sound recorders, and making and listening to music, are all spacetimemattering processes; the production of music, and the production of music-producing devices, are all apparatuses that create new material-discursive forms in the real world. A boundary as a spacetimemattering is something that happens according to the time of the experience. Boundary-making is time-making, it is a process of giving time to things; the listener, and the world in which they and the music they are listening to exist, are timed. In the same way that Olafur Eliasson's ice blocks impinge their non-human timeframes on their human audiences, music as a spacetimemattering force involves audiences and listeners in a timeframe that is not simply theirs to impose on music.

We can therefore understand the music on *Telyn Rawn* as the resultant pattern of the entanglement of Rhodri Davies, the telyn rawn and their respective 'spectral halos'. In this sense, Haraway's and Eno's images of composting can also be likened to such an accretion of meaning, where, in the context of Barad's quantum theory-inflected work, composting becomes superposition, and *Telyn Rawn* is a diffraction pattern as music. However, Barad's use of measurement, and measuring apparatus, enables us to think even more specifically about processes of making; making music, making harps, making meaning. Davies' own recreated telyn rawn and the music he creates with the instrument are points of measurement, of boundary-making and therefore meaning making. The harp, and the pieces Davies creates, are therefore a measuring, an ascertaining and a creative response to the interdependence of cultural and musical histories and heritages, the ongoing development of music through spontaneous acts of musical creativity and the simultaneously synthetic and authentic nature of new music. We can never truly understand and know what ancient music sounded like; we can merely make inferences and informed assumptions. In this sense, the historical enquiry of the *Telyn Rawn* project is a creative fabulation, but is no less authentic for that. As Barad suggests, all measurements are creative acts which position the measurer within the act of measuring. All measurements therefore leave a mark on a body; as creative director of the *Telyn Rawn* project, Davies is part of the reconstituted telyn rawn, as much as he is a part of the suite of pieces that are at once ancient and modern. As listeners, we hear all of this, and the pieces on *Telyn Rawn*, resulting as they do from the entanglement of wood, horse hair, cultural archaeology, and musical histories and futures, leave their marks on our bodies in the form of sonic impressions that are opaque and ultimately unknowable in their reality, and yet at the same time they are utterly transparent and accessible in their fusing of interdependent histories, practices and ideas.

In *The First Concert, An Adaptive Appraisal of a Meta Music*, the improvising musician, and drummer with the group AMM, Eddie Prevost discusses the way in which the practice of improvising music draws certain parallels with the notion of ecstasy. Following his own interpretation of the notion of 'ecstasy', Prevost states that, while improvising, musicians, 'loosen [their] own boundaries [and] exist within another dimension [...] I recall my own playing with early AMM. There was a common experience of being lost within the music' (Prevost, 2011: 87). For the writer and curator John Corbett, the experience of listening to a piece of improvised music can indeed be an equally disorienting process:

> Listening to [freely improvised] music in the moment is profoundly elastic [...] Without many of the usual markers, it's hard to know where you are in time, where you've been, and especially where you're headed, [it's less] a matter of not getting lost [but] an attempt at staying found.
>
> *(Corbett, 2016: 34–5)*

From both a musician's as well as a listener's perspective, these experiences also speak of Barad's notion of a mark being produced on a body. For Prevost and

Corbett, the 'mark' is the disorientation, the sensation of having an experience of music that is not tethered to either of their normal apprehensions of time. As suggested above, they are being pulled – as if by gravity – into the music's internal temporal flow; the music's time interferes with the musician's and the listener's time and a new sense of time emerges; an entanglement of music and musician, of music and listener. The marking of musical time does not solely reside in the music, the musician, the listener or the clock. It is in all of them at once. Thinking again of composition as sonic composting, we come to realise that time itself is part of the process of composting. Within Barad's quantum model, time therefore becomes another component of the diffraction/entanglement superposition, along with location, environment, musical instruments, listening equipment and so on, which do not simply affect the musician's or listener's experience, but fundamentally determine what is being produced. The act of listening is a time-based object – an event – just as a piece of music is, as is the process of making the music in the first place. Music as actually-existing thing-force, as a mark on a body, a boundary, a spacetimemattering, is not simply a relationship between a listener and a set of sounds. Instead, both music and listening are multiples, as are musician and listener. Please note, though: none of this is intended to confer special status on music. Just like everything else in the universe, music is fundamentally a spacetimemattering apparatus.

Ecomusicological Practices: Making Music and Making Musicians

The drummer Mark Guiliana has built a career that spans rock and pop music. Along with his own jazz and electronic music projects, he has extensive experience as a sideman, including his performances on David Bowie's final album *Black Star* in 2016. Discussing his favourite albums, Guiliana highlights the album *Drukqs* by Aphex Twin, the production alias of Richard D. James (Guiliana, 2020). Guiliana's interest and enjoyment in the album stems from James' approach to constructing the rhythmic elements of the album's tracks. As with another of Guiliana's album choices – *Feed Me Weird Things* by Squarepusher – James uses samples to create drum patterns which are then programmed in such a way that they simultaneously sound like a human drummer, and yet are playing rhythms that would be impossible for a human to play. Where Aphex Twin, Squarepusher and their contemporaries built on the affordances of sampling technology to create varieties of electronic dance music that evolved out of the drum'n'bass and jungle scenes of the 1990s that could be described as 'uncanny',[3] there are a variety of musical antecedents to this approach. For example, the player pianos of the late nineteenth and early twentieth centuries were loaded with rolls of paper, punched with holes arranged in patterns, that caused the piano keys to play themselves in an uncanny fashion, endlessly generating reams of music. In the age of driverless cars, we might think of these musical automata as player-less pianos. One their most notable proponents was the composer Conlon Nancarrow, whose *Studies for Player Piano* is an obvious antecedent for Aphex Twin's *Drukqs*; a series containing a number of pieces which – although rooted in blues and boogie-woogie styles – are exceedingly complex and impossible for any single human pianist to play.

Guiliana reflects on his own response to James' approach on *Drukqs*, saying,

> a lot of those things in regards to the drum set are technically impossible to achieve because it's not a drummer playing them and there might just be too much going on to actually pull it off in real time, but I did spend quite a bit of time trying to pull it off and failing happily, and I found that the ways in which I would fail would actually lead to some interesting results.
> *(Guiliana, 2020)*

In another interview, Guiliana relates that he was inspired by the rhythmic vocabulary created by producers like Aphex Twin and Squarepusher, given that, since they were often not drummers themselves, they were thinking differently about creating and programming drum patterns:

> I was inspired by that vocabulary, and tried to integrate it into my way of playing on an acoustic drum set. It's less a question of trying to recreate electronic music than of absorbing it as naturally as possible [...] I try not to think like a drummer, to change my perspective. I'm inspired by programmed beats, even if they are impossible to reproduce on a kit.
> *(Guiliana, 2018)*

The track 'One Month', from the Mark Guiliana Jazz Quartet album, *Family First* (2015), very much reflects Guiliana's interest in uncanny, programmed drum patterns. The introduction is a 24-bar passage in 4/4 time played initially by the drums and double bass – possibly a direct reference to the drum'n'bass stylings of the likes of Squarepusher – before being momentarily joined by the piano in bar 17. A crotchet pulse on the hi-hat remains constant throughout the introduction, while a highly syncopated interplay between the snare and the bass drum, which shifts between dotted quaver, quaver and semiquaver groupings of notes and rests, creates significant rhythmic tension and a frequent blurring of the metronome-like pulse, implying at times that the hi-hat has moved to the offbeat of the bar.[4] After the tenor saxophone enters at 46″, to play the main melodic theme of the tune, over a more conventional Latin beat, the drums and saxophone interlock in an improvised passage in A minor at 1′45″, and are joined at 2′42″ by the piano, moving the tonality to Bb minor. Throughout this section, there are echoes of Guiliana's interest in electronic music, in that his playing is both beat-like in the way that he uses the hi-hats, snare and bass drum – the three core components of conventional rock and hip-hop beats – to create patterns and improvisations, but all the while the underlying pulse of the track constantly moves around, implied rather than directly played. In other words, this section moves with a rhythmic flow, rather than a metronomic pulse. The instruments engage in complex interplay, the saxophone leading with melodic arcs, against which Guiliana's drumming picks out and emphasises certain phrases and propels the improvisation forwards with occasional references to the beat displacement patterns we heard in the introduction. Although 'One Month' contains many of the hallmarks of a jazz composition in its arrangement and sonic reference points, it is also full of Guiliana's deep investigations into the construction and constant destabilisation rhythmic patterns. The idea of playing drums as if he were not a drummer is a fascinating image of him trying to play against his intuition. Uncanny drumming; a drummer playing drums as if he were a machine playing drums, thinking like a drum machine would think with all of the flexibility of not being human: a drummerless drummer. Guiliana's playing on this track exemplifies the boundary-making effects of music, in ways that are directly redolent of Eshun's and Gilbert and Pearson's ideas. This is to say that Guiliana's music has gone beyond simply being an object of contemplation and appreciation. Through boundary-making, and the entanglement of his playing with the musics of Aphex Twin, Squarepusher and even Conlon Nancarrow, Guiliana's approach to drumming, and to thinking about and making music has been recalibrated. And further still, this is not simply a hearing of music on Guiliana's own terms, this is a diffraction of music and drummer that produces a certain way of making music.

A similar process of modifying performance technique can be discerned in the music of the improvising saxophonist John Butcher. In an interview with David Toop, Butcher speaks of his interest in improvisation as a process of discovery:

> I was drawn to those aspects of music where what you're engaged with is a mystery. Through being engaged with it you're trying to discover more about that mystery and particularly what lies beyond the horizon which you can't even glimpse yet. If you go through that process you hope to see that hidden part of the activity.
>
> *(Butcher, quoted in Toop, 2016: 174)*

This image of improvisation as a quest for something that he cannot recognise – a 'mystery' – in a place that he does not understand – 'beyond the horizon' – is not unlike Meillassoux's notion of the great outdoors. In other words, Butcher knows that something is out there; he just doesn't know what it is, or how to grasp hold of it. He goes on to say that,

> Very often in the course of rehearsing a gig something's happening in the music and you semi-hear in your head what you'd like to do but you don't know how to do it. Some of the time you will remember that when you're at home and start working on it. It doesn't come overnight, it's a series of very, very small discoveries that add up to something over a period. There was a time I got extremely methodical about it – for about a year of looking at all the possible fingering combinations and discovering the

> overtone spectrum and then finding which ones you could bring out multiple tones in that overtone spectrum.
>
> (*ibid.*: 174)

This sense of 'semi-hearing' is his musician's intuition for something on the periphery of his awareness, similar to Meillassoux's idea that we are able to recognise an outside of thought that we are not able to comprehend. As Butcher begins to play new tones in the overtone spectrum using new finger patterns, his playing is a searching for this new possibility that sits outside of his knowledge, his technique, his musical language. As we have already established, Meillassoux's brand of speculative realism does not allow for any kind of access to the absolute real, and as Alexander Galloway has said of François Laruelle's formulation of the One, Meillassoux's great outdoors is like encrypted data; we know it is there, but we cannot access it (Galloway, 2010). At its core, Meillassoux's work establishes a principle: while he does not provide us with a strategy, or a set of instructions about how to access an infinite beyond the confines of a presented world that we exist in, he does, however, suggest that logically, it must be possible to achieve such a going-beyond. We can choose which philosophical path to take – the ascetic, detached realism of Meillassoux and Laruelle, or the entangled materialism of Barad, Morton and Haraway – but Meillassoux's route offers an opportunity to reflect on how the things that we do are already instances of thinking beyond the confines of a correlation. Improvisation is just such an activity, demonstrating that, by virtue of its capacity to surprise both the improviser themselves as well as their audience, an improvisation is a reaching into the outside.

John Butcher's album *Invisible Ear* (Butcher, 2003) enables us to more readily imagine how a set of performances can emerge from the process of semi-hearing and searching beyond a barely visible horizon and make audible certain aspects of our philosophical enquiry. Four tracks are particularly suggestive of these ideas: 'Cup Anatomical', 'Streamers' and 'Bright Field'. To an extent, all four tracks exemplify the process of discovery during rehearsal that Butcher describes; a moment when a sound arrives that takes us by surprise, or that captivates because it was not quite what we expected. Sometimes such sounds can be the by-product of another sound that we are playing – an unwanted squeak that occurs whilst we are trying to play another note – and sometimes whilst we are just running through scales, warm-up exercises, or even just playing without any fixed point of focus. Equally, sounds and ideas can come from deliberately trying to play an instrument in a new, or at least a previously untried, way. The sounds on these pieces each speak of a moment of discovery or realisation on Butcher's part, of something happening that subsequently suggested further study, exploration and refinement. 'Cup Anatomical' is the sound of rapidly fluttering saxophone keys, as air is simultaneously blown through the instrument, raising and lowering the pitch of the airflow without sounding a note. At the same time, Butcher is using the sound of the bubbling spittle that gathers in the saxophone to create extra buzzing and whirring that is both reminiscent of fluttering insects and white noise. 'Streamers' is an improvisation that results from what could very well be the saxophone being held so close to a microphone that feedback begins to sound, but Butcher is careful not to step too far over this threshold, so that there is just enough feedback response to articulate the sound of the keys being lifted arrhythmically. As with Guiliana's interest in trying to not think like a drummer, Butcher's music here sounds like a saxophonist trying to exceed the limits of what a saxophone might be. He creates a metallic, percussive sound with his key presses, and as the track progresses he opens the feedback out more, in a see-sawing pattern that emerges between the percussive sound of the keys striking the saxophone's body and the hooting sound of the feedback response. The final track, 'Bright Field', explores the high-pitched squealing sounds that can result from overblowing into the saxophone's mouthpiece, and multiphonics, which creates two tones simultaneously.[5] In addition, Butcher uses other techniques, such as flutter-tonguing, which gives the sustained whistling sounds a sense that they are rapidly oscillating, as we hear with electrically produced feedback. These aspects of Butcher's work suggest that he is deliberately pushing the saxophone beyond its familiar limits, using the saxophone as an apparatus of discovery and mark-making, methodically identifying and shaping new sonic forms.

In a more recent set of improvisations with the guitarist Andy Moor, on the record *Experiments With A Leaf* (Butcher and Moor, 2015), we can hear that some of these earlier experiments have now resolved into techniques that are part of Butcher's sonic palette. During the track 'The Tongue Is A Flame', Butcher employs the feedback-sounding squeal that he explored during 'Bright Field' to counterpoint Moor's low, overdriven guitar sounds, and as Moor uses either a tremolo bar or the guitar's tuning peg to detune the guitar, we can hear Butcher splitting his single note into two intertwining whistles. Similarly, in the track 'Fantasy Downsize', from 4′30″ onwards, we hear a brief passage where Butcher makes use of percussive keying, to work alongside Moor's staccato string tapping. What the three *Invisible Ear* pieces comprehensively show is that improvisation is a process of coming to know that we don't know something, and of letting-in from outside of our understanding something that we did not completely design ourselves. This is not to say that the sounds themselves are in any way transcendental, or outside of an empirical framework – how can they be, they are simply the results of a set of physical interactions between mechanical and electronic sound-producing equipment and a human using their mouth, fingers and breath. Instead, we can hear that Butcher may not necessarily be driving all of these improvisatory processes and decisions himself; his entanglement with the saxophone as well as with his fellow musician Andy Moor, means that, as much as he is improvising music, music is improvising him.

Clearly, John Butcher is a musician who has very much invented the way that he plays the saxophone, and, while he may not be the only saxophonist to use this wider palette of sounds, he has undoubtedly created his own highly idiosyncratic musical language. In this sense we could say that Butcher's experiments with saxophone technique and sound production have not only produced a set of Butcher-like musics, but through this process he has gradually produced himself as a musician with an identifiable voice. A quote familiar to many musicians is this dictum attributed to Miles Davis: 'Sometimes you have to play a long time to be able to play like yourself' (Uitti, 2022). Reflecting on that line with a new materialist's ear, we can consider how the process of producing John Butcher or Mark Guiliana is not simply the result of 10,000 hours of practice; it is the entanglement with saxophones, drums, sound, other musicians, musical histories, sound technologies along with multiple other factors. All of this and more leads to the production of Butcher's and Guiliana's sonic imprints, of their instrumental techniques, of their musicality. In this sense, just as Guiliana has developed a recognisable approach in his rhythmic performance, and Butcher has blown through and pressed the keys of his saxophones, something has simultaneously pressed itself into their playing techniques, and the aesthetic choices that they are making, something that goes beyond their ability to simply author and control all of the sonic choices they have made.

Where John Butcher's performance practice focused on the affordances of amplified saxophone, Bendik Giske, across two albums – *Surrender* (2019) and *Cracks* (2021) – has developed an approach to solo saxophone performance that broadens out his field of material-discursive relations. In essence, he plays solo saxophone, making extensive use of two particular techniques: circular breathing and multiphonics.[6] In addition, Giske has worked with producers Amund Ulvestad (on *Surrender*) and André Bratten (on *Cracks*) to develop recording techniques that involve placing microphones in non-standard positions so as to capture a range of sounds that he produces whilst playing. For example, an array of microphones pick up the sound of the keys being depressed and released, which adds a rhythmic quality to Giske's music, and a contact microphone that is strapped to his throat captures the vocal sounds he makes when producing multiphonics. Along with recording, Giske makes use of this extended microphone array in live performance, as can be seen from a number of online videos, including in Matt Lambert's short film *Brist*, which sees Giske performing while walking through the empty HAU Hebbel am Ufer theatre in Berlin (Lambert, 2021). What can also be seen in Lambert's film is another one of Giske's key motivations that inform his music making: his perspectives on masculinity and queer cultures. In interviews in 2019, supporting the release of *Surrender*, Giske spoke of how an aspect of the album was his desire to explore 'the urgency to question masculinity' (Giske, in Covė-Mbede, 2019), and that 'circumstance and opportunity has led me to where I've found myself: enjoying club and drag culture, embracing the importance of taking a queer perspective, and playing saxophone' (Giske, in Giangregorio, 2019). The visual aspects of his performances are clearly a fundamental component of how he presents his work in the live domain, while at the same time, the performative nature of his compositions is equally fundamental to the experiences he sees himself creating for listeners: 'What I really offer in this album is my performance. The tracks are very stripped and each investigates very clear ideas' (Giske, in Covė-Mbede, 2019).

Before we turn to Giske's music itself, a final aspect that is key to understanding the depth of interdependence and ecologicality in Giske's work is his relationship to music and the performance of music as a physical process. Giske has acknowledged the impact that techno music has had on his development (ibid.), in particular, in club environments such as the Berghain nightclub in Berlin (Kuga, 2019). Writing for the music magazine *The Fader*, Mitchell Kuga relates that Giske's experiences of electronic dance music in a nightclub environment was critical to his creative development: 'It wasn't until Giske went to iconic Berlin nightclub Berghain that he reconciled his relationship with the saxophone' (ibid.). In addition, Giske acknowledges the influence of folk music on his approach to creating music, especially in regard to the relation between music and dance, where, for Giske, 'a lot of the rhythms, a lot of the sort of emphasis on pulse comes from the steps in a certain dance' (Giske, in Bola, 2022). Furthermore, Giske sees that music can be an 'iteration' of movement, another way of saying a development of music, rather than dance simply being a response to music:

> The saxophone is a wind instrument. Which means that it's very much a physical presence required. You have to reflect on your breath and develop breath techniques, and also embouchure techniques. I also use voice together with the saxophone and what I find is that there is both this materiality there, that doesn't adhere to just intonation that we're so tuned to in our ears. But also, it means that literally a hip movement has a large impact on my breath, and will generate a particular sound. Or it's a particular gesture in the wind that becomes audible [...] You could say it's music in a way but for me, it's also motion. It's a bodily movement. It's dance [...] it really is a way for me to kind of reflect on presence and movement and relation.
>
> *(Giske, in Bola, 2022)*

There is a clear sense in these perspectives on music and dance and the physical, material nature of each that, for Giske, they are simply different expressions of the same impulse: music as physical gesture, dance as a means to modulate and structure sound. Giske's music is rich with material-discursive entanglements and, for the listener, and more so for the audience watching a live performance, this is a music that is interdependent in every aspect; it is made of its interrelations, and Giske clearly strives to perform these interrelations in the live environment.

On 'Ass Drone', the first track on *Surrender*, we hear Giske exploring the space of the Vigeland mausoleum, and at times we hear the reflections of the saxophone's sound bouncing back from the walls. The track begins with Giske tonelessly blowing air through the saxophone, before introducing a low Bb (concert Ab), which remains as a constant fundamental drone through the piece. Initially, we hear this constant tone sporadically pulsing in volume, as Giske creates reverberations, possibly by moving around the mausoleum, finding different reflection points in the walls. At 1′10″, he gradually introduces a set of partials over the top of this note, using multiphonics to create an increasingly rasping sound. These begin with him emphasising the octave above the fundamental, which he slightly pitches up to imply a B natural at times, creating sonic tension. He also introduces an F natural (concert Eb) at times, a note that is a perfect fifth above the Bb, and therefore part of the latter's harmonic series. He swiftly cuts off his airflow at 3′18″, thereby terminating the sound and leaving the reverberations of the saxophone to gradually decay.

The second track on *Surrender* is 'Adjust', which was released as a stand-alone single in 2018, with an accompanying video (Giske, 2018). The video is noteworthy for two reasons with regard to the ecology of Giske's work. First, although the track – as with the music on both of Giske's albums – would not necessarily immediately register as a piece of 'pop' music, the fact that it had a music video made for it in some way places it within the sphere of popular music as something that can appeal to a non-specialist audience. Also, since Giske can be heard singing, almost keening, into the saxophone throughout 'Adjust', its emotive tone gives further weight to the track's pop sensibility. The second reason that the video is noteworthy is its contents. The video is a series of repeated slow-motion loops of a male dancer who is dressed only in a leather codpiece, a body chain and a rubberised head piece that combines the look of a motorcycle or space helmet – complete with a small light array that illuminates the dancer's face – with a set of long rubber tails, which he swings about like hair. The short loops are repeated moments of him dancing in a rapturous state to the track; the visual loops no doubt echoing the repeated phrases that comprise the musical itself. There is also a live video of Giske and the dancer – dressed in the same costume – appearing at CTM in 2019 (Giske, 2019b), a Berlin-based festival that presents contemporary electronic, digital and experimental music. In this video, Giske and the dancer move in time to a faster track, with the dancer placing emphasis on swinging his head around. The dancer's costume, and the sensual nature of the video, along with the festival appearance, all serve to indicate, as we have seen with Giske's own comments in his interviews, that his music comes from a place quite different from traditional, jazz-informed, solo saxophone. The dancer's costume appears as a visual reference to Giske's interest in queer cultures, and the enraptured dancing certainly signifies the kind of dancing that Giske associates with nightclubs such as Berghain.

The track itself is built on a simple repeating pattern in 7/8, a detail that quite possibly draws on Giske's interest in folk music. The seven beats per bar are in three rhythmic groups, 1–2, 1–2, 1–2–3, a pattern that is repeated over and over again throughout the piece.[7] We also hear clearly the other of Giske's interests: the closely miked keys of the saxophone, which create a rhythmic pulse throughout the track, and the use of singing at the same time as playing, to create multiphonic effects. Over the first ten seconds of the piece, Giske gradually introduces a two-note ostinato that repeats underneath the rhythmic pattern of the keys, initially moving between D and Bb, establishing the tonality as Bb major (concert Ab major). Having established these two foundational elements of the piece, at 1′00″, Giske starts to sing through the saxophone, a process which, because of the various constraints – singing through a small metal mouthpiece, recording the sound using a contact microphone strapped to his throat – produces the strained, keening tone that serves to heighten the emotional charge in his voice. The melodic phrase is simple, but extended, beginning with an F at 1′00″ (the third note of a Bb major triad, which is therefore implied by the accompanying D-Bb bassline underneath). It progresses to an E at 1′05″, at which point the bassline shifts to a C-Bb pattern, then onto a D at 1′20″, accompanied by a shift back to the D-Bb pattern, returning to E and C-Bb at 1′36″, before finally arriving back at F and D-Bb at 1′50″ (Giske, 2018).

With a change of producer, there is a marked change in sound for Giske's second album, *Cracks* (2021). The track 'Cruising' exemplifies this change, comprising a notable separation between the various sounds that Giske's performance produces – such as the key rhythms, the breath sounds, and even the different registers of the saxophone, each of which is refined in terms of its individual sound – at the same time as creating a deeply integrated overall sonic image for the piece. Cruising is in 4/4 time, a staple of club music, and as with 'Adjust', it is built on a repeated rhythmic pattern created by the amplified sound of Giske depressing the saxophone's keys. This is accompanied by a Bb minor-F minor arpeggio pattern, which André Bratten's production doubles in the early phase of the piece with a low octave, thus implying a club-like bassline. Throughout the track, Bratten's approach creates further connections to club and electronic music cultures, for example emphasising the bass and treble frequencies within the key presses themselves, which provides greater percussive propulsion, and from 2′30″ onwards, making extensive use of delay effects. Along with the technical innovations on *Cracks* that enabled Giske to further deepen and expand his creative approach, it is worth noting that the track's title, 'Cruising', can be taken as a direct allusion to the book *Cruising Utopia: The Then and There of Queer Futurity* by José Muñoz, a key reference point for Giske's conception of his creative music and sound practice (Rugoff, 2021 and Giske, in Bola, 2022).

Giske's approach encompasses a diverse range of ideas and practices, which in many ways completely disconnects him from the familiar canon of solo saxophone improvisers: Anthony Braxton, whose seminal recording *For Alto* (Braxton, 1971) drew together much of the learning that had happened in contemporary classical and free improvised music in the 1960s, along with exploring some of the more extreme sonic affordances of the saxophone itself; and Evan Parker, who pioneered the use of multiphonics and circular breathing in both ensemble and solo settings, but whose music bears almost no relation to the work produced by Giske on either *Surrender* or *Cracks*. Among more recent figures to emerge who have carved out a singular vision for the saxophone is John Butcher, who, as we have heard, has extensively explored the relationship between the acoustic and amplified sounds of a saxophone, using microphones to create feedback and other effects. Similarly, Colin Stetson has built a reputation for himself as a total saxophonist, creating muscular improvisations that use multi-phonics, circular breathing and microphone arrays, but in a way that is very different from that of Bendik Giske.[8] What's at stake here is that Giske's singular vision for saxophone-based music making is almost entirely outside a 'tradition' of improvised music making, that often references jazz, and a harmonic language that draws on jazz and contemporary classical structures. Instead, Giske sees his music as a 'contribution to a conversation [...] that happens in the realm of electronic music, clubbing, dancing and personal explorations. In that sense I really see my music as electronic music, even though it isn't electronic at all' (Giske, in Giangregorio, 2019). A final word on Giske's approach must come back to the importance of performance in his work. Just as Morton and Haraway repeatedly refer to the fact that the human biome is comprised of multiple non-human elements, so too are Giske and his music comprised of multiple musical and extra-musical forces. While Giske's detractors might seek to find fault in his work because of certain similarities to another saxophonist's music, pieces like 'Adjust' and 'Cruising' are made of dancing, breathing, movement, folk music, queer cultures and aesthetics, techno, nightclubs, as well as standard canons of jazz and saxophone techniques, along with Giske's own antipathies towards male-dominated, heterosexual music education systems (Giske, in Kuga, 2019).

All Listening Is Ecological

Writing in 2010, the composer Pauline Oliveros, known for her practice of 'deep listening' (Oliveros, 2005), further developed her approach with the concept 'quantum listening' (Oliveros, 2022). Oliveros' original conception of deep listening framed sound as a point of confluence of time and space, which she referred to as 'acoustic space' (Oliveros, 2005: xxiii). Where Truax and Schafer spoke of what we might think of as flat planes of sound, which is to say their models of acoustic communication and acoustic ecology are simply about categorising sounds, Oliveros recognised that sounds happen in temporal as well as physical planes. Oliveros defined deep listening as follows:

> Deep Listening for me is learning to expand the perception of sounds to include the whole space/time continuum of sound – encountering the vastness and complexities as much as possible. Simultaneously one ought to be able to target a sound or sequence of sounds as a focus within the space/time continuum and to perceive the detail or trajectory of the sound or sequence of sounds. Such focus should always return to, or be within the whole of the space/time continuum (context).
>
> *(ibid.)*

Thus, we can understand deep listening as a process of simultaneously listening to a wide field of sound, and recognising a range of sonic details that are occurring in time and space – an approach not unlike acoustic ecology – while at the same time focusing in on the details of a particular sound or sounds. There are creative as well as ethical concerns here. As an improvising musician, Oliveros was developing a way of thinking about sound at multiple scales, so that a musician might be more consequent about the sounds they choose to make during an improvisation. In deep listening mode, an improvising musician could be listening to the entire field of sound that they are immersed in – for example, all of the sounds in a concert hall, or all of the sounds in an outdoor location if they are playing outside. They could be listening to the sounds of the ensemble they are playing with as a gestalt whole, and at the same time they could be listening to individual sounds in the performance environment and/or individual sounds and notes being produced by the other musicians that are around them. All the while, they would be listening closely to the sounds that they themselves are producing, and being attentive to the quality of these sounds, and the relationship between the sounds that they are making and the various fields of sound they are interacting with. Thus, deep listening is a process of listening to multiple scales of sound at once, while scaling up and down to consider sonic qualities in both stand-alone and relational contexts.

Oliveros also compares deep listening to meditation, describing how it is intended to heighten one's consciousness of sound as well as the process of hearing itself; what she refers to as 'attentional dynamics'. In this sense, she relates that 'deep listening comes from noticing my listening or listening to my listening' (ibid.) This process of tuning into and focusing on the act of listening is about being attentive, but it also connects to a sense that listening is a creative act; listening listens to things in a certain way, and puts things into certain relations with other things. Listening is therefore not simply something that just happens, the listener happens it. The more recent essay 'Quantum Listening' takes this line of reasoning further:

> Quantum Listening is listening in all sense modes to or for the least possible differences in any component part of a form or process while perceiving the whole and sensing change. The Quantum Listener listens to listening. Quantum Listening simultaneously creates and changes what is perceived. The perceiver and the perceived co-create through the listening effect.
>
> *(Oliveros, 2022: 52)*

Although she does not articulate it as such, Oliveros' concept of quantum listening implies, alongside Barad, an acknowledgement of Bohr's and Schrödinger's work on superposition and observation. There is a sense in her ideas that quantum listening is about recognising the simultaneity of sonic events, that listening – as a process of measuring – momentarily brings sonic superpositions into points of experience. Ultimately, it would seem that the 'quantum' in quantum listening is a reference to the role that measurement plays in deciding the nature of an outcome; listening is thus not simply a cool, detached observation of a process that is unfolding at arm's length, it is a constituent part of the process itself. Thus, within the context of 'quantum' listening, 'listening to listening' becomes a process of acknowledging – beyond what Oliveros was exploring with deep listening – the part the listener plays in creating the sonic world that they are in. Importantly, this is not simply correlationism by another route. Quantum listening is couched in terms of the physical and temporal properties of sound, in relation to the attention of the listener, and we are presented with a mode of listening that bears comparison to agential realism. Thus, where Bohr's experiment showed that the process of observation brings electrons into view as particles, quantum listening listens to all sounds in a given context and acknowledges the role that the listener has in giving those sounds certain form. In this sense we can also identify the ethical imperatives in Oliveros' deep/quantum listening practices, but we can go further than this to think of Oliveros' ideas as a composting of Schafer's work on acoustic ecology. As with the Strathern/Haraway perspective that it matters what listening we choose to listen with, and Morton's reflection on how *Ice Watch* demonstrates the folly of trying to impose human time on a really-existing world of non-human creatures and forces, quantum listening is a call to make considered choices about the sounds we produce, but also to understand the non-neutrality of the choices we make about how we listen.

In *The Nuclear Sonic: Listening to Millennial Matter*, Lendl Barcelos discusses and compares the work of the sound artist Jacob Kirkegaard and Peter Cusack, whom he describes as a 'sonic journalist' (Barcelos, 2016: 71). Both Cusack and Kirkegaard had created audio projects that were based on sound recordings made within the zone of exclusion at the Chernobyl nuclear plant in Ukraine, site of the historic reactor meltdown on 26 April 1986.

For the album *4 Rooms*, Kirkegaard made a set of recordings in four different spaces at the abandoned Chernobyl site; a church, an auditorium, a swimming pool and a gymnasium. Subsequent to the initial recording in each space, Kirkegaard then implemented a layering process first used by the composer and sound artist Alvin Lucier, in his landmark work *I Am Sitting In A Room* (Lucier, 1990). The piece is based on a recording of Lucier reading out a passage of text. The text begins, 'I am sitting in a room, different from the one you are in now. I am recording the sound of my speaking voice and I am going to play it back into the room…' (Lucier, 1990), before it goes on to detail how the piece is going to be constructed, how it will sound and, in the language of acoustics, why it will sound that way. The recording is then played back into the room, while simultaneously being recorded itself. That recording is then played back into the room, while it too is recorded. That recording is then played back into the room, while being recorded, and thus the piece progresses.[9] As the process repeats, the sound becomes increasingly indistinct; as

Lucier relates in his speech, the blurring of his voice is the result of his voice being overlaid by the sonic characteristics of the room itself. Along with being a conceptual piece of sonic art and music, Lucier's piece is a simple demonstration of how sounds interact with the resonant frequencies of physical spaces. Kirkegaard's *4 Rooms* is very much an extended exploration of this principle, a meditation on the process of listening to listening in the context of a zone of nuclear fallout.[10] Across the four pieces, Kirkegaard removes any human input – which is to say that he does not create any new sounds in each of the four rooms himself. This is simply the sound of a recording device listening to a room, and then listening to that recording, and repeating that process over and over again. This is inhuman quantum listening, what Barcelos describes as a dehumanising process (Barcelos, 2016: 74). Kirkegaard's recordings of recordings are in themselves diffraction patterns, reminding us that the human ear, while it is certainly the fulcrum point that produces the agential cut, is by no means the only thing in the room that is listening.

Barcelos contrasts Kirkegaard's approach with that taken by Peter Cusack in the *Sounds from Dangerous Places* project (2011). For Cusack, 'sonic journalism is based on the idea that all sound, including non-speech, gives information about places and events and that listening provides valuable insights different from, but complementary to, visual images and language' (Cusack, 2022). Within this description, there are a number of key themes that Cusack's work explores. The first is that sonic journalism works to de-centre speech. In traditional journalism, the journalist reports the story, using sounds and images to bring their report to life. In sonic journalism, speech is only one aspect of a sound recording; it is not being used to tell the story, or to convey key facts and messages, it is simply part of the sonic image. Second, Cusack is interested in how contextual information can enhance the experience of listening to field recordings, which he compares to the use of captions and titles being used to deepen the experience of looking at a photograph (Cusack, 2022). Third, and connected to this last point, Cusack suggests that field recordings 'convey far more than basic facts' (ibid.), listing a sense of spatiality, atmosphere and timing, along with emotional content as some of the valuable insights that sound offers, distinct from visual images and language (ibid.). Released in 2011 as a double CD package of recordings from Azerbaijan, Chernobyl, Kazakhstan and various UK sites (including Dungeness, Sellafield and Snowdonia), along with an accompanying book of photographs, interviews and other documentation, *Sounds from Dangerous Places* brings together these three vectors to convey in sound the complexities of the geopolitical forces that have shaped, and continue to shape, these environments. Listening to three tracks on the *Sounds from Dangerous Places* project website – 'Cuckoo By Radiometer', 'Chernobyl and Power Cable Crackle' and 'Chernobyl' (Cusack, 2022) – I am immediately struck by the sonic complexities to which Cusack alludes. The contrast between the sounds of the radiometer and the cuckoo, and the buzzing cable against the backdrop of birdsong is stark. And yet, as he suggests, there is an emotional veracity to these recordings that goes beyond being a simple condemnation of the environmental degradation caused by the reactor breakdown. The recordings convey a sense of space, but also conjure sonic relations between a Geiger counter and a cuckoo that stem from their rhythmic interplay, particularly from the fact that their frequencies and patterns are in some way harmonically aligned in the key of D major (the cuckoo's repeated 'cu-coo' is an F sharp to D refrain, while the radiometer arrhythmically 'chirps' what approximates to a high E natural). Thus, 'Cuckoo By Radiometer' is a curious sonic assemblage: the combination of cuckoo and radiometer tells of the devastation caused to natural environments by human dereliction, while at the same time, the Geiger counter–cuckoo duet is a serendipitous musical occasion. In this context, listening to Cusack's work is also an example of quantum listening, as the piece sets up a number of vectors that inform what we might focus our listening on, not least the sense that we are listening to ourselves listening; in other words, listening to how we listen, and listening to the choices we make about what we are listening to.

For Barcelos, what aligns Cusack's and Kirkegaard's work is that each of them uses sound recording to in some way capture, and certainly convey, a sense of non-human time through sound; what he terms 'the nuclear sonic' (Barcelos, 2016: 72). Barcelos himself draws on Kodwo Eshun's ideas on how 'millennial durations' (Eshun, in Barcelos, 2016: 85) can be given affective, material expression; in other words, using the human timescale of a sound recording to convey the half-life decay times of radioactive isotopes such as strontium, cesium and plutonium, all of which have been found in the ground at Chernobyl. According to Ukrainian authorities, it could take up to 320 years for the area around the reactor to become habitable (Madrigal, 2009 and Nijjar, 2011), while Greenpeace has reported that 'areas close to Chernobyl NPP may be uninhabitable for tens of thousands of years' (Dawe et al., 2016). Thus, quantum listening, in the context of the nuclear sonic, is as much a listening to non-human time as it is to cuckoos and radiometers. In this way, Barcelos' insights offer a way to further refine the temporal aspects of Oliveros' concepts of deep/quantum listening. Barcelos' focus on the presence of radioactive half-life time within human-centred time in *4 Rooms* as well as *Sounds from Dangerous Places*, simultaneously reminds us that non-human time is as much a part of the world in which we exist as cuckoos and swimming pools, and that listening, as a process of measuring, is a process of creating time. The nuclear sonic, quantum listening, sonic fictions and agential realism all direct us to the conclusion that listening is an ecological process; ecological listening thus emerges both as a practice – we can notice that we are listening ecologically – and more importantly, as an ecological process: all listening is ecological.

Music-fiction

Across this chapter, we have listened to and looked at how musicians have entangled themselves in interdependent webs of ideas, technologies, histories, techniques, instruments, identities and different scales of time in order to create music that is simultaneously of these interdependencies, and yet completely of itself. To complete this mattering of music, I want to now turn to a theory of practice, of creative production, to further extend ways of thinking about how we make music, and how music makes the world. Across two books, *The Concept of Non-Photography* and *Photo-Fiction, A Non-Standard Aesthetics*, François Laruelle formulated the term 'photo-fiction', a concept that activates his theories of non-philosophy in the context of creative practice. The word fiction, and the resultant concept 'fictioning', name a particular approach to understanding creative practice, within the context of Laruelle's non-philosophical mode of thought.

The language of non-philosophy operates in a very particular way, and among his key terms, Laruelle establishes the One as a universal which constitutes the whole of reality. To recap, everything that happens, or exists, is within – or given in – the One, an absolute Real that is immanent to the world that we experience all the time. Just like Meillassoux's conceptual configuration of an absolute real that is 'outside' of thought, Laruelle's Real is not a product of human thought. We cannot experience it, conceive of what it is, and we cannot knowingly reproduce it. However, although the Real is foreclosed to human knowledge, as with the 'great outdoors', we are able to acknowledge its existence. Laruelle's Real is not within us, we are within it, and whenever we make something, interact with others, or have any kind of experience, we are able to acknowledge their concrete and objective reality because they exist in the Real. Unlike the outdoors of Meillassoux's speculations, which is used to demonstrate the concrete existence of an absolute reality beyond our capacity to imagine it, the Real is the ground for everything that we experience; the world is produced within the Real. Being 'given in' the Real means that we are both created by the Real, and always a part of it.

In terms of creative practice, one of Laruelle's most compelling formulations is the idea that, through acts of creativity, we humans are able to produce the world. This is because, being 'given in' the Real means that our actions are not only in the world of observable experience, but they are simultaneously part of the inaccessible Real. This is to say that everything is 'real' at the same time as being 'Real', which means that everything we do has a double aspect; it is R/real. Laruelle uses the term 'photographic stance' to describe how this process of simultaneous production happens in photography. In the process of making a photograph, a photographer is making something that becomes part of the real world of experience, and at the same time they produce the Real, the conditions of the world (Laruelle, 2012: 12–13). Extending the idea of 'cloning' that we encountered in the previous chapter, Laruelle calls this 'performing' the Real as a way of thinking more directly about creative practices. He refers to the photographic stance as a type of performance because, while it is a way of understanding how creative acts are processes of creating the world, within his vision-in-One model, humans are not able to access, conceive of, or understand this absolute reality of which we are a part. All we can do is mimic, clone or *perform* the process of creative production, in this case, by taking a photograph.

Performing the Real is therefore a process of doing without knowing, and of making without understanding, which Laruelle articulates as a process of producing something that is 'parallel' to the Real (Laruelle, 2011: 19).

Making things is therefore a performance of the Real, and the photographic stance is just such a performance, where world-making operates according to the processes and contexts of photography. Taking a photograph is a process that takes place in the real world of tangible experience and at the same time is a performance of the inaccessible Real. As a result, a 'photo-fiction' – essentially a photograph that exists in the world of lived experience – is both a tangible object that we can apprehend, and something we can never understand.[11] The photograph's tangibility makes it a parallel of the Real. While it in no way *represents* the Real, the act of producing the photograph is nevertheless an act of producing the world.

Alighting on Laruelle's phrase 'Principle of Sufficient Photography', David Burrows and Simon O'Sullivan further emphasise the importance of this strange dual aspect of what we could call the photographic R/real. Using the word 'truth' to underline the non-subjective nature of a photograph (in other words, a photograph's capacity to index something that is not a property of our interaction with the photograph), they suggest that photography's grounding in the Real means that its claims to truth are not based on its capacity to represent the Real. Rather than representing the Real in the world of real experience, the photographic stance has a claim to truth precisely because it suspends phenomenological experience and interpretation of experiences in the world. The photograph is true, not because it is of the real world, but because it is of the Real (Burrows and O'Sullivan, 2019: 324–5). Extrapolating from Laruelle's concept of photo-fiction, the process of *fictioning* – which means 'producing truth' – is grounded on the idea that a photo-fiction '"pictures" what happens to experience when not tied to a self/interpreter, or, again, when such experience is not "processed" through representation' (ibid.: 325). For Laruelle, a photograph as photo-fiction is non-representational. The objects within a photograph function within the context of *that* photograph – they are R/real because they are 'in-photo' – and are not grounded by their capacity to relate to or represent anything outside of the photograph. Given that photography is fundamentally concerned with producing images of other things, Laruelle's position seems curious. However, as we have seen with new materialist and speculative realist perspectives, Laruelle's non-representational model is built on the principle that the things we experience in the world are objectively real, not because of their (cor-)relation with each other, but because of a non-subjective, absolute Real, which by virtue of being immanent to the world of lived experience, grounds everything in that world.

We can apply Laruelle's ideas about the performative process of adopting a creative 'stance' to music. A musician has an experience of their instrument, of their given performance environment, of their own thoughts and responses to that environment, for example, their interplay with other musicians, the sonic properties of a performance space, and the sounds that they themselves are producing on their instrument. Just as the contents of a photograph are in-photo, these sounds, patterns and structures are in-music. The choices about what sound to make, what kind of musical structures could be developed, what kind of instrumental technique should be used, all form aspects of the music stance. Furthermore, in the process of creating music, everything is happening according to music.

Alexander Galloway's discussion of the concept of fiction and what he calls the 'death of representation' in aesthetics and creative practice further emphasises commonalities between Laruelle's non-philosophical project and the trajectories of speculative realism and new materialism. Galloway suggests that this death has occurred in both new materialism and in what he calls 'the end of critique', a stance that reminds us of Meillassoux's disdain for 'post-critical thought'. Boundary-making as non-representational creative practice is certainly a valuable tool for thinking about how things are made and how this creative process produces meaning. That is to say that a boundary, as the point that gives definition to aesthetic experience, does not represent anything beyond itself. Even in the context of thinking interdependently, it could be argued that Bendik Giske's music is not representing the manifold elements that shaped its development. Instead, at the point of performance, the music is nothing but itself, even though it was formed through Giske's entanglement with many practices and discourses. Galloway extends this idea further, and considers performance, invention, creativity, artifice and construction (Galloway, 2014: 158) in relation to fiction as outlined above. He goes beyond seeing representation as a process of representing other things in the world around us, and instead positions creativity-as-fictioning as non-representational. Galloway's point is that, although experience is a process that happens in the world, experience is itself something that is given in the One, and because being given *in* the One is not an experience *of* the One, it follows that experience is non-representational. By this logic, in creating music-fictions, musicians are producing parallels of the Real. Galloway calls this process 'radical objectivity', suggesting that fiction 'is neither a fictionalised version of something else, nor does it try to fabricate a fictitious world or narrative based on real or fantastical events' (ibid.). This sense that fiction produces an objectivity without representation to an extent is akin to both Haraway's and Barad's views about the meaning-making processes of boundary formation; the meaning produced by boundaries is not a representation of the things that come together to produce the boundary, instead, the boundary is the meaning.

Although Laruelle's ideas appear to be a radical departure from the other ideas we have encountered in this chapter, on closer inspection the similarities may be more compelling than the differences. If music is non-conceptual, as well as being both non-representational and non-expressive of anything other than itself, there are some resonances with theories of boundary-making, spacetimemattering, and indeed, subscendence. This is to say that music, or anything that results from a creative process – art, computer programming, garden design, cookery – while it can be made of, and made with, any number of things, this is not the same as saying that art represents or expresses those things. Composted vegetable matter might be made of kitchen waste, but it is compost, not an expression of kitchen waste. The boundaries-as-aesthetics produced by the entanglement of Bendik Giske's circular breathing technique, use of Balkan folk rhythms, interests in queer cultures, and resonant frequencies in the Vigeland mausoleum are the aesthetics of the track 'Adjust', they are not simply representations of any of those things. The same can be said of Chiu and Honer's *Recordings from the Åland Islands*, of Davies' *Telyn Rawn*, and indeed, of all of the interdependent music that we have listened to throughout this chapter. In Timothy Morton's words, these musics are 'subscendant'; they are all less than the sum of their parts.

Conclusion: Music as an Ecological Real

Throughout this chapter, I have reflected on various pieces of music, and approaches to listening and making that enable us to hear and think about processes of meaning making. While the musical examples I have explored are particularly transparent, and lend themselves to ecological ways of hearing and listening with an ear to how the interdependent nature of music gives voice to the interplay of its component parts, these pieces and processes are by no means unique. As I hope to have shown, music is ecological and interdependent not only because it is made out of all available materials, materials which are themselves ecologies of discursive-material practices, but because it also forms new ecologies with the discursive-material practices that it encounters. Thinking ecomusicologically about practices of making and listening to music is to understand these practices as boundary-making and mark-making practices. These are practices that draw together histories, cultures, techniques, technologies, not simply to make something that is of these forces, but to make something with them, something that is different, separate, distinct and new. Donna Haraway's ideas can frame music as a pattern that is pattern forming, and about composition as composting. While a composer might also be a composter, they are also composted by the materials they use to make music. Although we make music, music also makes us.

Where Naess saw ecological enmeshment as something separate from the real world, what Morton, Barad, and even Meillassoux and Laruelle to an extent, are working to articulate is an ecological real. Importantly, these configurations suggest that music is a site for the production of real experience and meaning, that meaning can be both produced and shared through music, and that practices of music making and listening have the capacity to create real meaning in the world that is grounded in a non-subjective, absolute real. These new materialist, speculative realist, non-philosophical and deep ecology perspectives can therefore lead us to an appreciation of listening and music making that does not separate these practices from the real. Making music and listening to it are not simply subjective intuitions, they are concrete engagements with the real. Music as an ecological real is not only made up of all the things in the world, it also consists of, and is constitutive of, an absolute real; a real that we constitute and which constitutes us, and at the same time, it exceeds our capacity to conceive it. Making and listening to music is the agential cut that we make in order to both join with, and create, the world around us.

Notes

1 For example, the band Stick in the Wheel ran the *Perspectives on Tradition* project, which enabled contemporary musicians to explore the folk music library and audio archive at Cecil Sharp House, examining the meaning of, and their relationships with, traditional music and culture (Stick in the Wheel, 2022).

2 The Emanuel Vigeland Museum website describes the mausoleum as a 'dark, barrel-vaulted room, completely covered with fresco paintings [...] The impression of the dimly lit frescoes with multitudes of naked figures is reinforced by the unusual and overwhelming acoustics of the room' (Emanuel Vigeland Museum, 2022).

3 In his 1919 essay, 'The Uncanny', Freud makes reference to the psychiatrist Ernst Jentsch's original formulation of the concept of the uncanny. For Jentsch, the uncanny takes the form of doubt as to whether a lifeless object may in fact be alive, or something we think is alive turns out to be inanimate, citing the disturbance we may feel in the presence of waxworks, life-like dolls and automata (Freud, 2003: 135). Following this notion, it could be said that some of the musical productions of Aphex Twin, Squarepusher and their ilk are uncanny, in that the kinetic rhythm patterns that they create often have a sound that is remarkably human-like (since many jungle and drum'n'bass tracks were created using samples of recordings that featured human drummers), and yet utterly impossible to play, given the detailed editing and complex structuring that is the hallmark of these artists' work.

4 It is also worth noting that the rhythmic introduction theme of 'One Month' is itself a reformulation of rhythmic ideas that appear in a more electronic arrangement – including a sampled typewriter – on the track 'Locked in a Basement' (2006) by one of Guiliana's earlier groups, Heernt.

5 Multiphonics – literally 'many voices' – is a technique used across various instrumental families, whereby the player creates more than one tone from a monophonic, single-voiced source, for example woodwind and brass instruments, and the human voice. To create the effect on a saxophone, a player can sing into the instrument, rather than simply blowing, which causes the saxophone to produce two notes at once, or they can hold the instrument's keys down in certain combinations, which can also produce two or three simultaneous notes.

6 Circular breathing is a technique practised in several music cultures across the world, frequently associated with the didgeridoo, but widely used across Asia; indeed, Giske notes that his first exposure to circular breathing was as a child in Bali, learning to play the flute. The technique consists of creating a constant, smooth flow of air out of the mouth, seamlessly moving back and forth between breathing out and using the cheeks as bellows to push out air whilst simultaneously breathing in. Many of Giske's pieces last for around ten minutes and consist of an unbroken flow of sound that he produces by using the circular breathing technique.

7 In Balkan traditions these groupings are referred to as 'quick' (1–2) and 'slow' (1–2–3), and feature in many traditional dances. For example, the dance *Eleno Mome*

(Smithsonian Folkways Recordings, 1957) would be notated 'Q,Q,S' (Folkdance Footnotes, 2021).

8 Stetson has made extensive use of an array of techniques (including circular breathing and multiphonics) on a series of albums since his debut solo release *A New History Warfare, Vol. 1* (2008). It is interesting to note that in the comments section of the 'Adjust' video on YouTube, viewers are keen to point out the similarities between Giske's and Stetson's music. Indeed, Ekte Boi writes 'Am I missing sth or this guy just a Colin Stetson rip-off? Seems like the exact technique and aesthetic,' to which Jake Burgess replies 'total rip off' (Giske, 2018). While there are certain sonic similarities between Giske's and Stetson's approaches, to reduce Giske's saxophone playing and compositional approach to a bare comparison of stylistic details seems to miss the broader discourses in which Giske's work resides, particularly his interests in foregrounding his own connections with electronic music, nightclubs and queer cultures as exemplified in the music video for 'Adjust' itself.

9 On the recording issued by Lovely Music, the speech is repeated and recorded 32 times, 45′21″ (Lovely Music, 1990).

10 It is also worth noting that Bendik Giske interacts extensively with the resonant frequencies of the Vigeland mausoleum on the recordings that comprise the album *Surrender* (Giske, 2019).

11 Among a wave of publications and responses to Laruelle's non-philosophy project, Gunkel, Hameed and O'Sullivan's *Futures & Fictions* (2017) and Burrows and O'Sullivan's *Fictioning* (2019) have engaged with the opportunities offered by Laruelle's conceptualisation of fiction, exploring it in the context of a range of critical and creative practices.

Bibliography

Amgueddfa Cymru. 2022. *Folk Songs*. Available at https://museum.wales/collections/folksongs/ (accessed August 2022).

Barad, K. 2007. *Meeting the Universe Halfway*. Durham and London: Duke University Press.

Barcelos, L. 2016. 'The nuclear sonic: Listening to millennial matter,' in Brits, B., Gibson, P. and Ireland, A. (eds), *Aesthetics After Finitude*. Melbourne: re.press.

Bola, D. 2022. 'A conversation with Bendik Giske'. Available at https://weare-europe.eu/a-conversation-with-bendik-giske/ (accessed August 2022).

Braxton, A. 1971. *For Alto*. Chicago, IL: Delmark Records.

Burrows, D. and O'Sullivan, S. 2019. *Fictioning: The Myth-Functions of Contemporary Art and Philosophy*. Edinburgh: Edinburgh University Press.

Butcher, J. 2003. *Invisible Ear*. Italy: Fringes Recordings.

Butcher, J. and Moor, A. 2015. *Experiments With A Leaf*. Amsterdam: Unsounds.

Chiu, J. and Honer, M.S. 2022. *Recordings from the Åland Islands International Anthem*. Available at https://intlanthem.bandcamp.com/album/recordings-from-the-land-islands.

Clera. 2021. *The Harp*. Available at http://www.clera.org/saesneg/harp.php (accessed August 2022).

Corbett, J. 2016. *A Listener's Guide to Free Improvisation*. Chicago and London: The University of Chicago Press.

Cové-Mbede, H. 2019. 'Bendik Giske – Breath, body and metal: fearless beauty'. Available at https://metalmagazine.eu/en/post/interview/bendik-giske (accessed August 2022).

Cusack, P. 2011. *Sounds from Dangerous Places*. ReR MEGACORP.

Cusack, P. 2022. *Sounds from Dangerous Places*. Available at https://sounds-from-dangerous-places.org (accessed August 2022).

Davies, R. 2020. *Telyn Rawn*. Amgen.

Davies, R. 2022. *Rhodri Davies*. Available at http://www.rhodridavies.com (accessed August 2022).

Dawe, A., McKeating, J., Labunska, I., Schulz, N., Stensil, S.-P. and Teule, R. 2016. 'Nuclear scars: The lasting legacies of Chernobyl and Fukushima'. Available at https://wayback.archive-it.org/9650/20200409062940/http://p3-raw.greenpeace.org/international/Global/international/publications/nuclear/2016/Nuclear_Scars.pdf (accessed August 2022).

EFDSS. 2022. *The English Folk Dance and Song Society, Cecil Sharp House*. Available at https://www.efdss.org (accessed August 2022).

Emanuel Vigeland Museum. 2022. 'The museum'. Available at https://www.emanuelvigeland.museum.no/museum.htm (accessed August 2022).

Eno, B. [1982]1986. *Ambient 4 (On Land)*. Sleeve notes (revised 1986). New York: E.G. Music.

Eshun, K. 1998. *More Brilliant than the Sun*. London: Quartet Books.

Folkdance Footnotes. 2021. 'Eleno Mome (L*), Elenino Horo – Bulgaria'. Available at https://folkdancefootnotes.org/dance/a-real-folk-dance-what-is-it/about/eleno-mome-elenino-horo-bulgaria/ (accessed August 2022).

Freud, S. 2003. *The Uncanny*. London: Penguin Books.

Galloway, A.R. 2010. *French Theory Today: An Introduction to Possible Futures A pamphlet series documenting the weeklong seminar by Alexander R. Galloway at the Public School New York in 2010. Pamphlet 4 | Quentin Meillassoux, or The Great Outdoors*. Available at http://cultureandcommunication.org/galloway/FTT/French-Theory-Today.pdf (accessed July 2022).

Galloway, A.R. 2014. *Laruelle: Against the Digital*. Minneapolis, MN: University of Minnesota Press.

Giangregorio, T. 2019. 'Bendik Giske: Q&A'. Available at https://flaunt.com/content/bendik-giske?rq=bendik (accessed August 2022).

Gilbert, J. and Pearson, E. 1999. *Discographies: Dance, Music, Culture and the Politics of Sound*. London: Routledge.

Giske, B. 2018. 'Adjust '. Available at https://www.youtube.com/watch?v=j8OyappCpxw (accessed August 2022).

Giske, B. 2019. *Surrender*. Oslo, Norway: Smalltown Supersound.

Giske, B. 2019b. *Bendik Giske – Live at CTM 2019*. Available at https://www.youtube.com/watch?v=L00__csoJRg (accessed August 2022).

Giske, B. 2021. *Cracks*. Oslo, Norway: Smalltown Supersound.

Guiliana, M., 2018. In Scheinin, R. 'Jazztronic: Mark Guiliana finds fertile ground between Jazz and Beat Music'. Available at https://www.sfjazz.org/onthecorner/mark-guiliana-jazztronic/ (accessed August 2022).

Guiliana, M. 2020. *'My Top 10' Albums With Drummer/Bandleader Mark Guiliana*. Available at https://www.youtube.com/watch?v=oaTLVk-eD-8 (accessed August 2022).

Gunkel, H., Hameed, A. and O'Sullivan, S. 2017. *Futures & Fictions*. London: Repeater Books.

Haraway, D.J. 2016. *Staying with the Trouble: Making Kin in the Chthulucene*. Durham and London: Duke University Press.

Heernt. 2006. *Locked in a Basement*. New York: Sunnyside.

Ice Watch. 2015. *Ice Watch: An Artwork By Olafur Eliasson and Minik Rosing On the Occasion of COP 21 – United Nations Conference on Climate Change*. Available at https://icewatchparis.com (accessed August 2022).

Joyce, C. 2018. *Makaya McCraven's Utopian Vision of Jazz Could Change the World.* Available at https://www.vice.com/en/article/yw9y3v/makaya-mccraven-universa l-beings-noisey-next-interview (accessed August 2022).
Kerson, S. 2019. 'Makaya McCraven: The brain behind the mind-bending beats'. Available at https://www.npr.org/2019/02/27/698697855/makaya-mccraven-the-brain-be hind-the-mind-bending-beats (accessed August 2022).
Kuga, M. 2019. 'How Bendik Giske deconstructed himself through saxophone play'. Available at https://www.thefader.com/2019/02/08/how-bendik-giske-deconstructed-himself-through-saxophone-play (accessed August 2022).
Lambert, M. 2021. *Brist.* Available at https://www.youtube.com/watch?v=QRtWvJp yy3M (accessed August 2022).
Laruelle, F. 2010. *Philosophie Non-Standard: Generique, Quantique, Philo-Fiction.* Paris: KIME.
Laruelle, F. 2011. 'The generic as predicate and constant: Non-philosophy and materialism', in Bryant, L., Srnicek, N. and Harman, G. (eds), *The Speculative Turn: Continental Materialism and Realism.* Melbourne: re.press.
Laruelle, F. 2012. *The Concept of Non-Photography.* New York: Urbanomic.
Lovely Music. 1990. *Alvin Lucier: I Am Sitting In A Room.* Available at http://www.lovely.com/titles/cd1013.html (accessed August 2022).
Lucier, A. 1990. *I Am Sitting In A Room.* New York: Lovely Music.
Madrigal, A. 2009. 'Chernobyl exclusion zone radioactive longer than expected'. Available at https://www.wired.com/2009/12/chernobyl-soil/ (accessed August 2022).
Mark Guiliana Jazz Quartet. 2015. *Family First.* Bahamas: Fifty-Six Hope Road Music.
McCraven, M. 2018. *Universal Beings.* Chicago, IL: International Anthem Recording Company.
McCraven, M. 2018b. *Where We Come From (Chicago x London Mixtape).* Chicago, IL: International Anthem Recording Company.
McCraven, M. 2022. *Makaya McCraven: Beat Scientist, Drummer and Producer.* Available at https://www.makayamccraven.com (accessed August 2022).
Morton, T. 2017. 'Subscendence'. Available at https://www.e-flux.com/journal/85/ 156375/subscendence/ (accessed August 2022).
Morton, T. 2018. *Being Ecological.* London: Penguin Books.
Nijjar, R. 2011. 'Chornobyl by the numbers'. Available at https://www.cbc.ca/news/ world/chornobyl-by-the-numbers-1.1097000 (accessed August 2022).
Oliveros, P. 2005. *Deep Listening: A Composer's Sound Practice.* Bloomington, IN: iUniverse.
Oliveros, P. 2022. *Quantum Listening.* Newcastle upon Tyne: Ignota Books.
Prevost, E. 2011. *The First Concert: An Adaptive Appraisal of a Meta Music.* Copula.
Rugoff, L. 2021. 'Norwegian saxophonist Bendik Giske announces LP on Smalltown Supersound'. Available at https://thevinylfactory.com/news/bendik-giske-announces-new-album-cracks-vinyl-smalltown-supersound/ (accessed August 2022).
Sherburne, P. 2022. 'Snåcko: This single from the Los Angeles multi–instrumentalist duo's forthcoming album evokes the stillness of a remote Scandinavian island'. Available at https://pitchfork.com/reviews/tracks/jeremiah-chiu-marta-sofia-honer-sna cko/ (accessed August 2022).
Smithsonian Folkways Recordings. 1957. *Eleno Mome.* Available at https://folkways.si.edu/the-dances-of-the-worlds-peoples-vol-1-dances-of-the-balkans-and-near-east/music/a lbum/smithsonian (accessed August 2022).
Stick in the Wheel. 2022. *Perspectives on Tradition.* Available at https://stickinthewheel.bandcamp.com/album/perspectives-on-tradition (accessed August 2022).

Sutherland, T. 2016. 'Art, philosophy and non-standard aesthetics', in Brits, B., Gibson, P. and Ireland, A. (eds), *Aesthetics After Finitude*. Melbourne: re.press.

Toop, D. 2016. *Into the Maelstrom: Music, Improvisation and the Dream of Freedom*. London and New York: Bloomsbury.

Uitti, J. 2022. 'The 15 best Miles Davis quotes'. Available at https://americansongwriter.com/the-15-best-miles-davis-quotes/ (accessed August 2022).

Vaughan Williams, R. 1991. *Serenade to Music, The Lark Ascending, Fantasia on Greensleeves, English Folk Song Suite, In the Fen Country, Norfolk Rhapsody No.1*. London: EMI Classics.

Vaughan Williams, R. 2008. *Folksong Arrangements*. London: EMI Classics.

3

MUSIC, RIGHTS AND REVENUE

In the previous chapter, we listened to music in the context of materialist-realist perspectives and took an ecological approach to thinking about music, and the experiences of producing and listening to music. As we saw, the philosophy does not make music ecological; ecomusicological listening is an approach that recognises music's ecologicality, foregrounding the entangled and interdependent material-discursive – rather than correlational – nature of the relations between music, its constituent parts, music makers and listeners. The following two chapters take up this theme of entangled interdependency, tracking the proliferation of music within two digital environments, and exploring how two different ecosystems operate in relation to music production, modes of consumption and practices of listening. In this chapter, we engage with music in the context of traditional rights-based models of distribution, with a particular focus on music's relation with two techno-economic paradigms that have emerged, and rapidly evolved in recent years: blockchain technology and song catalogues. In the next chapter, we turn to music's increasingly entangled relationship with social media platforms, and the rise of the creator economy.

In November 2020, at the Digital, Culture, Media and Sport Committee's Inquiry into the Economics of Music Streaming, the MP Damian Green asked Guy Garvey, lead singer of the group Elbow and BBC6 Music DJ, 'does the dominance of streaming now influence the way you write music or perform it? Does it have a direct effect on the artistry?' (Green, UK Parliament, 2020). Garvey's response was that the group had 'altered the intro to a song for fear that it would not be included on playlists. We had a long introduction to the first song on our last album, which we clipped for streaming use' (Garvey, UK Parliament, 2020). Later in the interview Garvey widened his perspective on the impact of streaming of musicians' creativity, saying, 'I think it is a problem that

DOI: 10.4324/9781003225836-4

continuing exploration sonically, creatively, has come to an end because [the streaming] business model does not suit it' (ibid.). Clearly, Garvey speaks from a particular perspective, and across the enquiry his responses largely reflected his own experiences and views about how creative music making can be best facilitated within the rock and pop music sector, where artists make use of commercial recording studios to create, record and produce their work. However, despite the particularities of Garvey's responses, they are an indication of how technological and economic changes within the music sector have impacted on the creative outputs of musicians' – and listeners' – habits and preferences.

In what follows, I explore the part that recorded music's current format as an intangible data asset is playing in catalysing change in music production and consumption habits, focusing on how rights and revenue patterns are evolving. In the previous chapter, we centred on constructing a framework to understand the ecological nature of music. Now, by investigating how emergent technological and financial services are offering new revenue models for music creators and rights holders, the focus is on bringing an interdependent perspective to the creative practice of sustaining a career as a musician.

I begin by reflecting on the work of the sociologist Mary Mellor (Mellor, 2019) and the economists Jonathan Haskel and Stian Westlake (2018), and Raghuram Rajan (2019). Their work indicates how value systems, market dynamics, financial products, and indeed money itself in the early twenty-first century, have changed, and how, within a systems framework, these changes are connected to broader social, political and ethical patterns. The chapter then builds on this foundation to examine two commercial models that have opened up new business structures within the digital music economy: music and song funds, and blockchain-based streaming platforms. Expanding out from a decade-long incubation in fintech, blockchain-powered streaming platforms, rights management systems, along with coherent strategies for implementing collaborative and cooperative-based digital organisations, have rapidly proliferated in recent years. The result is a holistic view of digital music commerce, one that acknowledges the interdependent interplay between the specific changes caused by technological development, in the form of emergent commercial platforms and blockchain technology, and wider market economics and value systems.

Since beginning the research process that underpins this chapter, the two trajectories that I map out have continued to develop at pace. While it is not my intention to provide an exhaustive overview of the arc of development within either music and blockchain cultures or the music catalogue economy – that would require a substantial focus in each case, beyond the scope of a single chapter – it is nonetheless both valuable, as well as fascinating, to observe these fast-evolving trajectories. My aim has been to examine these two areas and to identify and document how these developing contexts for music have created opportunities for music practitioners, demonstrating in a broader ecological sense the interrelations between changing technological paradigms and changing practices in music.

In the book *Network Science* (2016) Albert-László Barabási considers different types of network: cellular networks (consisting of the interactions between genes, proteins and metabolites), neural networks, social networks, communications networks, along with power grids and trade networks. While these networks are composed of different materials – living and non-living, organic and inorganic – for Barabási, the structures underpinning these networks share fundamental laws and organising principles (Barabási, 2016). Barabási goes on to propose that our networked world/s are essentially composed of complex systems, which expresses the idea that it is difficult to identify a linear relationship between a system's components and their aggregated behaviour (ibid.). The overarching narrative in this chapter tracks certain developments in blockchain technology and the recent and rapid expansion of the song catalogue market, and considers these in terms of entangled networks of digital technology and creative production. As a result, I am interested to show how music's interplay with evolving techno-economic paradigms has become both visible and audible in contemporary practices of creative music production.

A Hybridised Music Industry

Understanding the changing nature of digital music commerce, and recognising how the sector is adapting to technological change, is fundamental to building a holistic and systems-based view of music. Therefore, before engaging with wider perspectives on how value and financial systems are changing, it is important to establish why these matters are important to a contemporary understanding of music.

> Platforms are not just software. They involve the 'stacking' of hardware, software, goods and services, and interfaces into a user experience / marketplace ultimately controlled by one entity.
>
> *(Meier and Manzerolle, 2018)*

In 'Rising Tides? Data Capture, Platform, and New Monopolies in the Digital Music Economy', Leslie Meier and Vincent Manzerolle (2018) offer a series of case studies and insights into the increasingly symbiotic relationship between the music and information industries. Previous shifts in the music economy have seen numerous format changes, the assimilation of traditional record companies into media conglomerates and a trend towards monopolisation, which has resulted in the current hegemony of the 'big three' multinationals (Sony BMG, Universal Music Group and Warner Music Group). However, Meier and Manzerolle's elaboration of the elision of music commerce with the IT industry enables them to articulate the emergence of substantially new business models and behaviours within what they refer to as the 'digital enclosure' economy. Their analysis of both Jay-Z's collaboration with Samsung on the release of his 2013 album, *Magna Carta...Holy Grail*, and Apple's 2014 marketing campaign which featured the automatic installation of U2's *Songs of Innocence* 'directly into the library of 500 million iTunes subscribers' (Sherwin, 2014), demonstrates how current practices have engaged in platform commerce in a way that integrates conventional download models with the harvesting of user data (Meier and Manzerolle, 2018). The article also shows how the data-led disruption of music commerce has been building for several years, within both the download and streaming markets; suggesting that developments in digital commerce regularly precipitate more general changes in both content production and consumption habits. In more precise terms, they propose that the transformation of music from physical to digital artefact 'highlights the links between marketing, user data, and the use of music as data' (ibid.). This hybridisation, bringing together entertainment and metadata, is a clear indication of the ecological nature of the contemporary digital music economy. At a fundamental level, the way that music operates as a shared experience within digitally enabled environments has been altered by its integration and entanglement with a range of extra-musical technologies, processes and dynamics.

Meier and Manzerolle's analysis of the release of the *Magna Carta* album examines the extent to which the project was an exercise in 'data mining' (ibid.). They proposed that the project revealed that 'the strings attached to free music [are] user data' and their analysis demonstrated how the opportunity to listen to music was 'exchanged for compromised user privacy and forced, automated, word-of-mouth promotion via social media' (ibid.). In addition, and perhaps more significantly in terms of the evolution of the digital music environment, they saw how the album's release 'offered a decisive step towards embedding music within a platform logic driven by IT interests, testing data-capture-based business models' (ibid.). As such, their work clearly describes how a fundamental system change – that is, the evolution of recorded music to digital information artefact – has led to the tethering of music production and distribution to data capture and analysis processes on an industrial scale.

The article goes on to discuss Drake's *More Life* playlist release in 2017, describing it as a 'promotional and data event' that combined 'the star- and hit-driven economics of the music industries and the logics of platform accumulation' (ibid.). The power of this message is that it details the emergence of a new form of music commerce, one embedded within a digital data framework. We might call this *datafied* music commerce, rather than simple online sales. However, it would seem that musical artists are yet to see the benefits of this hybrid model, and the article suggests that what we might call a 'streaming-as-data mining' distribution and market research strategy primarily operates to shore-up and strengthen income streams and profit-making opportunities for large media and technology companies. Fundamentally, the greatest impact of platformisation of the digital commons has been an ever-growing asymmetry of wealth distribution across the music sector. In one of his contributions to contributions to the UK Parliamentary enquiry, the Radiohead guitarist Ed O'Brien expressed a similar view, saying 'the way that streaming services pay out benefits, the big artists get millions and millions and millions of streams. Yes, it creates a sort of bias within the system' (O'Brien, UK Parliament, 2020).

Music Creators' Earnings in the Digital Era (Hesmondhalgh et al., 2021) is a UK Government report published in 2021 that provides an in-depth analysis of streaming economics, detailing how royalty payments are calculated and allocated to rights holders, and an expansive overview of the changes that have occurred to industry revenues during the rise of music streaming platforms. The report's main focus is on the changes that streaming has had on music creators' capacity to generate income, and earn a living from the commercial distribution of their music, which includes physical sales, downloading, sync licensing, public performance and broadcast in addition to on-demand streaming. It defines 'music creators' as 'performers'; in other words, the vocalists, MCs and instrumentalists who perform for recordings, the composers and songwriters whose works are created in order to be recorded, and the producers and engineers whose creative input is integrated in recordings in a variety of ways (ibid.). It is important to distinguish this use of the term 'creator' from its use within the context of the 'creator economy', which we shall examine in the next chapter. The authors note that the global value of music copyrights has increased significantly in recent years, along with 'a considerable increase in the amount of revenue coming into the music industries since 2015, as a result of the rise of streaming' (ibid.). However, their analysis shows that, in comparison with 2001 – which they term the 'peak CD' year – current revenue levels for the recording industry are now significantly lower. Exacerbating this drop is a notable increase in the number of musicians aiming to generate income from their music. The report draws on the analysis of former Spotify Chief Economist Will Page who shows that, since the launch of Spotify in 2009, the number of UK songwriters has increased to 140,000 (a rise of 115 per cent), and the number of UK recording artists has risen to 115,000 (an increase of 145 per cent) (Page, in Hesmondhalgh et al., 2021). In simple terms, more and more creators are now competing for a share of a smaller overall sum. Within the falling overall figure, the report identifies two revenue streams: recording revenue and publishing revenue. While recording revenue decreased from the late 1990s through to the early 2010s and has not fully recovered, publishing revenue has continued to grow (ibid.). The report also provides some sobering figures for the proportion of music creators achieving significant numbers of streams, and the thresholds required to make a sustainable living from streaming. The top 10 per cent of artists account for 98 per cent of all streams, the top 1 per cent for 78–80 per cent of streams, and the top 0.1 per cent for 39–43 per cent of streams, where there has been a small, but recognisable decline since 2016. For those wishing to make a living from streaming, then the report suggests that 1 million streams per month are required to achieve this, which, for UK artists, would include both UK and non-UK streams (ibid.). Overall, Hesmondhalgh et al. see that it is the so-called creative 'middle class' – the 200,000 artists who, according to Spotify's Loud&Clear data portal, generate between

$1,000 and $1 mn in payments for rights holders – which requires the most attention in terms of assessing the extent to which streaming can facilitate a sustainable livelihood, and whether or not streaming has eroded the 'musical middle class' (ibid.).[1]

Evolving Money and Evolving Value

Beyond catalysing radical change in music commerce, the hybridisation of the digital music economy enables us to think ecologically about the creation and use of money, offering new perspectives on how we understand money and its relation to value. Reciprocally, as part of an evolving ecology of music, changing conceptions of money and value are themselves further evidence of music's interdependent nature. For example, although the 'Rising Tides' article and the *Music Creators' Earnings in the Digital Era* report suggest that, for many musicians, creating material change in their personal circumstances is frequently elusive. Nonetheless opportunities to generate revenue from music are becoming more prevalent, with the result that the scope to identify as a professional musician – a 'creator' – is increasing.

In her mapping of the origins of contemporary exchange systems, Mary Mellor informs us that 'money is a social and political construct' (Mellor, 2019: 69), whose 'roots are in society, not the market' (ibid.: 48). Mellor's project is to assert that while it may be tempting to think that money has value because of some intrinsic property (such as precious metal scarcity), instead, we think of money as a 'vital social institution' precisely because it does *not* have an inherent value (ibid.: 153). In this, Mellor views money not as a 'thing', but as 'a network of promises and obligations' (ibid.). Given its functional basis in trust and obligation, she also accounts for how money is practically generated, identifying state spending and bank lending as the two most obvious examples of how new money comes into being. During late 2008 and early 2009, in response to the global financial crisis, Mellor describes how governments across the world used state spending, quantitative easing and low interest rates as means to generate new money, feeding it into both the financial and wider economic sectors, and argues for greater involvement of citizens and workers' groups in the monitoring of public spending (ibid.: 150). As a result, we can think of money as an ecosystem comprising the financial strategies of the state and banking sector, integrating with 'social trust and public authority' (ibid.: 81).

In *The Third Pillar* (2019), Raghuram Rajan's approach is similarly ecological, positioning the state, markets and the community as the three necessary components of a successful and thriving economy. As with Mellor, Rajan identifies an imbalance towards the private sector, and a lack of overarching awareness within core assumptions about value creation and the function of wealth distribution. We can make a comparison between Rajan's approach and the work of the economist Elinor Ostrom, who, in her work on the management of Common Pool Resources (for example, water and fisheries), showed that such shared resources could effectively be managed by 'polycentric institutions'. Ostrom saw these as a combination of governmental and non-governmental actors, working together to achieve 'effective, equitable, and sustainable outcomes' (Ostrom, 2010: 664). As with Ostrom, Rajan's approach is to see that reconciling the goals of the public and private sectors is central to developing more equitable and sustainable markets (Rajan, 2019: 370). Thus, Rajan understands that it is necessary for firms to go beyond their traditional model of seeking to maximise financial return for shareholders, and instead look to increase value for a range of stakeholders, including employees and suppliers, via a process he refers to as 'firm value maximisation' (ibid.: 372).[2] Rajan contrasts firm value maximisation with a corporate management approach that prioritises 'profit maximisation'. He cites the work of 20th century economist Milton Friedman who saw that 'management would maximise economic value by maximising the value of the equity' (ibid., p371), an approach reflected in the general trends outlined above, where the platformisation of music streaming is seen as having largely benefitted technology and media companies, rather than musical content creators. In firm value maximisation, however, 'firm value' is defined as 'more than just the value of equity and debt investments in the firm' and includes 'the value of specific investments made in the firm by those who have a long-term attachment to it' (ibid.: 372). Whether we think in terms of 'firm value maximisation' or 'participatory budgeting' (Mellor, 2019: 150), what comes through in Rajan's and Mellor's work is an imperative to go beyond historical financial precedents, and to find more equitable and sustainable approaches to macroeconomic development. These perspectives also serve as a reminder of how the digital evolution of both money and value are informed by, and inform, the interaction and consensus of a wide set of stakeholders, and, for our current purposes, given the digital nature of the vast majority of commercially traded music, Mellor's insights into hold here too.

The economist Mariana Mazzucato adds to this perspective in her analysis of state support of the commercial research and development sector, with a particular focus on innovation-led growth. Where traditional perspectives on public support of the private sector might assume that this would involve 'subsidies, tax reductions, carbon pricing and technical standards' (Mazzucato, 2018: 207), Mazzucato shows that innovation, and therefore innovation-led growth, is not solely the product of state-sponsored research and development programmes. Instead, again reflecting Ostrom's conception of polycentric institutions and common resource management, Mazzucato suggests that innovation comes about through institutional partnerships which 'allow new knowledge to diffuse throughout the economy' (ibid.). In essence, it is the interdependent relations between the public and private sectors, and between different institutional actors within these sectors that enable meaningful development and innovation to occur. Critically, Mazzucato describes her approach as a 'systems perspective' (ibid.: 210), using this term to emphasise the complex, and nonlinear, nature of a state's relationship with innovation-led growth. Mazzucato's particular focus is on developing an infrastructure for innovation; one which goes beyond simply funding innovation.

Although the three economists address these matters in different ways, there is a discernible consensus around a core set of values and ideas. Their systems-based approaches to understanding economic planning and stimulus, and their promotion of infrastructural change as a means to facilitate participatory and multi-institution collaboration across all aspects of the public and private sectors, offer further perspectives on thinking ecologically about music, and its dynamic interrelations with a global digital economy, and with state and corporate finance. Mazzucato, writing for the *Harvard Business Review* in 2013, highlighted that both Apple and Alphabet (still trading as Google at that time), had both been beneficiaries of significant government funding for technology and algorithm development (Mazzucato, 2013). Such deep ecological connections between the digital music environment of streaming platforms and social media, and public and private sector interests, is a clear – although easy to miss – indication of how policy decisions, public and corporate funding, and a variety of polycentric institutional agendas are part of the fabric of the curation and recommendation processes, which in themselves have given shape to current practices of listening to, sharing, creating and making a living from music.

Intangible Economics

Beyond the wider systems-led economic analysis and calls for change, both Mellor and Rajan recognise the role that digital technology is playing in the transformation of some of the more operational aspects of commerce that are relevant to music production and distribution. Rajan lists 'information, knowledge, creative works, and ideas' as the key components of the evolving data economy (Rajan, 2019: 378), reflecting the insights that Meier and Manzerolle developed in relation to digital music. In an extended analysis of intellectual property, Rajan discusses advances in the data economy, particularly with regard to 'intangible assets', describing a song as a 'nonrival', intangible asset, since one person listening to a song does not prevent anyone else from listening to it (ibid.). Given the continued focus on interdependence in this chapter, it is also notable that Rajan positions intellectual property in relation to the information economy. He asserts that music has value because of its embeddedness in a wider set of legal and cultural structures, and suggests that, 'if a song could be sung by anybody, the songwriter could never benefit monetarily from her creativity; without legal protection, intellectual property, especially property that needs to be used publicly, would have no value' (ibid.). Echoing Mazzucato's analysis, Rajan goes on to argue that all intellectual property – again, including music – 'gets its value from past innovation and government protection rather than necessarily the owner's continued innovativeness or efficiency' (ibid.). This highlights the reliance the private sector has on the public sector in terms of both legal and digital frameworks. Given this relationship, Rajan proposes that regulations relating to patents should be examined and modified in line with the evolution of the data economy, reflecting Mellor's ideas about participatory monitoring of public spending.

In *Capitalism Without Capital* (2018), an extensive analysis of the emergence of intangible assets, and the implications of the ongoing growth of the 'intangible economy', the economists Jonathan Haskel and Stian Westlake similarly examine the evolution of intellectual property, patents and copyright. Their analysis offers further definitions of intangible assets in terms of ideas, knowledge and social relations (Haskel and Westlake, 2018: 4), and includes designs, recipes, brands, licensing agreements, design skills, the MP3 protocol and music rights as examples of this type of asset (ibid.: 10–11, 60). There are two key messages that we can take from Haskel and Westlake's research. The first is that the ongoing evolution of intangibles has resulted in new markets for music rights within equities markets, which I discuss below. The second is that, within the context of the data economy, intangible assets are creating increasingly diverse opportunities to control personal data and to generate copyright revenue. It is this second insight that forms the basis of the later discussion of blockchain streaming platforms, and I examine use cases where blockchain technology has been used to explore new opportunities for rights protection and revenue generation from digital music formats.

Catalogue and Song Funds

Haskel and Westlake (2018) discuss what they refer to as 'superstar markets', where technology, for example broadcast media, has greatly increased the reach of superstar workers – including popular music artists and premier league football players – leading to huge increases in their earnings potential. The analysis identifies two opportunities for superstar owners of intangibles. In the first instance, Haskel and Westlake suggest that 'tech billionaires' stand to benefit hugely from the equity stakes they own in the companies they have founded. Second, they propose that superstar asset owners have 'special privileges to create more of a certain type of intangible', citing the fact that only J. K. Rowling can write new Harry Potter novels as an example (ibid.: 130–1). Their findings and conclusions about the development of intangible assets within equities markets reflect the emergence of a new asset class that has responded to the digitisation and concurrent financialisation of music, one based on songwriters' catalogues. As the music journalist and music business analyst David Turner has noted in his *Penny Fractions* newsletter, in the years leading up to 2010, a series of private equity firms began to acquire a number of the music industry's largest firms (Turner, 2020). The music journalist Gabriel Meier expands on the point to describe how this activity also reflected a 'strategic shift', in that private equity firms moved away from acquiring major labels, and began to focus instead on purchasing artists' song catalogues and – at times – individual songs (Meier, 2022).

Hipgnosis Songs Fund described itself as 'the first UK investment company offering investors a pure-play exposure to Songs and associated musical intellectual property rights' (Hipgnosis Songs Fund Limited, 2020). Launched on the FTSE 250 London Stock Exchange in 2018, its founder Merck Mercuriadis had raised interest in the company by suggesting that songs were a safer investment than either oil or gold, since legacy songs were something that people turned to in both good times and bad. Speaking in 2021, he proposed that, unlike gold or oil, whose price could be affected by global events (such as the global COVID-19 pandemic or Russia's invasion of Ukraine), 'proven' songs (his example was the Eurythmics' 'Sweet Dreams (Are Made of This)'), were 'part of the fabric of society' (Mercuriadis, 2021). This is to say that, as far as Mercuriadis is concerned, these proven songs, and the income generated from them, are not vulnerable to global events, since they resonate with matters of personal significance and nostalgia. From its launch onwards, Hipgnosis' focus was on building a portfolio of songs written by 'globally important' songwriters which it sees as having 'cultural importance' (Hipgnosis Songs Fund Limited, 2020). In its 2018 IPO (initial public offering), the fund raised £200 mn and subsequently went on to raise over £1.05 bn in a series of issues through 2019 and 2020 (Music Business Worldwide, 2021a). In early 2021, Hipgnosis Songs Fund acquired 1,180 compositions, half of the worldwide copyright and income interests in Neil Young's song catalogue, for an estimated $150 mn (approx. £109 mn along with all of Fleetwood Mac guitarist and singer-songwriter, Lindsey Buckingham's publishing catalogue (Ingham, 2021). The fund also owns the entire song catalogue of numerous legacy artists, including Justin Bieber, Barry Manilow, Chris Cornell, Blondie, Chrissie Hynde, Kevin Godley and many others (Hipgnosis Songs Fund Limited, 2020). In simple terms, Hipgnosis' goal has been to secure an income for its shareholders by investing in songs and the intellectual property rights associated with them; aiming to own '100% of a songwriter's copyright interest in each song, which would comprise their writer's share, their publisher's share and their performance rights' (Hipgnosis, 2021). Writing in the *Financial Times* in early 2021, Jamie Powell suggested that Hipgnosis' approach to investing in songs had been facilitated by the 'recurring and growing nature of payments from streaming platforms', which, unlike the unpredictable sales patterns of physical formats, was more akin to the behaviour of bonds (Powell, 2021). In October 2022, Hipgnosis announced a debt-funded share buyback scheme after its share price fell by 30 per cent over a six-week period. The fall in share value meant that Hipgnosis' market value was at half the level the company had said its catalogues were worth. At that point, the *Financial Times* described Hipgnosis' growth as having happened at 'breakneck pace, as it repeatedly raised new money to purchase more catalogues' (Wiggins and Nicolaou, 2022). The report showed that, as of October 2022, Hipgnosis had not bought a catalogue for over a year, and the revenues generated by its portfolio of songs had been falling, and that its falling share price meant that it was unable to raise new equity to purchase more catalogues without risking negative impacts for Hipgnosis' shareholders (ibid.). However, in early 2023, in a deal purported to be worth approximately $200 mn (£161 mn), Hipgnosis Songs Capital announced that it had bought shares of Justin Bieber's publishing and recorded-music catalogue, along with 100 per cent of Bieber's interest in his publishing copyrights, master recordings and neighbouring rights for the entire Justin Bieber back catalogue, totalling more than 290 titles released before 31 December 2021 (Aswad, 2023). With Bieber's songs having been streamed over 32 bn times on Spotify, Mercuriadis re-emphasised the significance of the singer's impact on global culture across his 14-year career, and of the deal itself, ranking as one of 'the biggest deals ever made for an artist under the age of 70' (Mercuriadis, in Stassen, 2023). While the financial challenges of 2022 and Hipgnosis' restructuring as Hipgnosis Song Management – an overarching company that manages the investments of the US-based Hipgnosis Songs Capital and UK-based Hipgnosis Songs Fund – speak of a volatility in the songs fund sector, this latest deal suggests that for Mercuriadis and the Hipgnosis portfolio, the financial opportunities associated with the exploitation of legacy artists' catalogues remain considerable.

Founded in 2010, Round Hill Music progressed through a series of multi-million-dollar funding rounds before Round Hill Music Royalty Fund Ltd launched its own IPO on the London Stock Exchange in November 2020, raising a further $282 mn (approximately £205 mn). Yet another public launch in December 2020 raised a further $46.1 mn (approximately £34 mn. These public launches enabled Round Hill Music Royalty Fund to acquire songs by The Beatles, Celine Dion, Louis Armstrong and the Rolling Stones (Ingham, 2020a), in addition to its ownership of Carlin Music, which included the Elvis Presley catalogue (Music Business Worldwide, 2021b). In its Investment Policy, Round Hill Music Royalty Fund stated that its strategy has been to invest in the publisher's share, writer's share and performance rights of a songwriter's copyright interest, along with the master recording rights of the recording of a song (Round Hill Music Royalty Fund Limited, 2021). The fund also confirms that its intended investment targets are classic legacy copyrights, which already exhibit a proven appeal to listeners and which 'are not subject to the natural decline in earnings and value that typically occurs within the initial ten years of a composition's life' (ibid.).

In December 2020, the *Wall Street Journal* reported that the music publisher Primary Wave had acquired 80 per cent of the Fleetwood Mac singer and songwriter Stevie Nicks' publishing catalogue, valued at approximately $100 mn (£73 mn) (Steele, 2020). This latest acquisition added to the interests that Primary Wave already owned in the catalogues of Ray Charles, Bob Marley, Whitney Houston, Burt Bacharach, Smokey Robinson and the Four Seasons, among others (Ingham, 2020b). Having raised $1.7 bn (£1.4 bn) through private equity in October 2022 to support further investment in music rights (Stassen, 2022), it is clear that for Primary Wave, alongside Hipgnosis Songs Fund and Round Hill Music Royalty Fund, its two main challengers in the catalogue fund market, is continuing a trajectory that pursues acquisition of legacy artists is a high priority.

As a newly lucrative and increasingly accessible class of intangible assets, the rapid rise of these catalogue funds signals a consequential change in how income is being generated from music-based intellectual property. While many legacy artists and rightsholders may be happy to go against a decades-long tradition of collecting royalties against their publishing and performance rights, and surrender their claims to the highest bidder, there are also many artists and musicians aiming to earn a living via the royalty payments system. It is this latter group who will find it increasingly hard to compete in a digital music ecosystem that is aligned with both the supply side and demand side dynamics of the legacy market.

In the superstar equity markets that Haskel and Westlake discuss, such moves to create and manage payment streams by 'aggregating catalogues of songs into one fund' (Powell, 2021) have developed new revenue streams for creators, which are not reliant on traditional revenues from royalty payments. Reporting on Primary Wave's acquisition of the royalties of the band Air Supply, MusicAlly confirmed that, as a result, Primary Wave stood to gain from the success of the band's royalties, rather than from receiving payments relating to the band's publishing rights as such (MusicAlly, 2020). Similarly, Round Hill Music Royalty Fund Limited, which described itself as 'a private equity firm exclusively dedicated to investments in revenue generating music copyright assets', reported that it has 'collected over US$175 mn of Net Publisher's Share on behalf of funds it manages' (Round Hill Music Royalty Fund Limited, 2021). The important issue here is not the amounts of money involved, but the financial mechanisms that the likes of Hipgnosis, Round Hill Music and Primary Wave are using. Meier frames the rise of song funds in terms of the wider financialisation of the music sector, and frames increasing 'marketisation and assetisation of royalties and individual song rights' – in other words, a complete shift to a digital music economy that operates according to the market logic for intangible assets – as the cause of a fundamental shift in how music is valued (Meier, 2022). Within this, Meier highlights how, even at the level of measuring or predicting an artist's commercial performance within the intangible music economy, companies such as Royalty Exchange are moving away from traditional performance metrics, such as 'an artist's ability to sell records, MP3s, streams or concert tickets' (ibid.), towards metrics that are native to the music equities environment, including 'Dollar Age'.[3]

Legacy Music, Listening Habits and Artists

In terms of our overarching ecological framework, a consequence of the emergence of investment markets for songs and associated songwriters' rights, is that funds such as Hipgnosis Songs Fund and Round Hill Music Royalty Fund are creating new opportunities to monetise music through financial products.

However, rather than simply creating revenue via traditional systems that capture sales, broadcast figures or even streams, it is the growing capacity of music to function as a tradable asset on the stock market that is generating this value. This new-found relationship between music and financial markets demonstrates a revaluing of music, and to an extent a revaluing of distribution mechanisms, such as streaming platforms. It is notable that, at this stage of music's development as an intangible asset, legacy songs are finding most traction. Given artists such as Neil Young and Barry Manilow's already-secure reputation as songwriters and the proven cultural value of their songs, this is unsurprising, but the success of already successful music as an asset sends a signal to emergent artists, in terms of the widespread value that is apportioned to music by the listening public.

Writing in 2022, Meier reported that in 2021 the private equity company The Blackstone Group had entered into a $1 bn partnership with Hipgnosis Songs fund and private equity firm KKR to purchase Kobalt Capital's music rights portfolio of 25,000 songwriters and 600 publishers, including Diplo, Paul McCartney and The Weeknd (Meier, 2022). In 2017, Blackstone had acquired SESAC, the performance rights organisation that licenses the public performance of the catalogues of 30,000 songwriters, comprising over 1 million songs (SESAC, 2022). For David Turner, Blackstone's influence on the current financialisation of the music sector has been pivotal, citing their advice to Sony Music to purchase CBS Records in 1986 as a major milestone in the development of the current music equities ecology (Turner, 2020).

According to the music historian Ted Gioia's analysis of Luminate's music industry data,[4] in 2021, the share of catalogue songs – defined here as songs that are over 12 months old – being listened to across streaming platforms rose from 65.1 per cent in 2020 to reach 69.8 per cent (Gioia, 2022). In consequence, the share of new music being listened to fell from 34.9 per cent to 30.2 per cent. This is reflected by Hesmondhalgh et al., who note that, in 2015, in the top 0.1 per cent of the most popular music, the ratio of new to catalogue music was largely equal, while, since 2018, the proportion of new to catalogue music has been in sharp decline (Hesmondhalgh et al., 2021). The number of catalogue albums being streamed in 2021 was 623.6 million compared with new album streams of 269.5 million (Gioia, 2022). It is clear that 'proven' catalogue music has taken over from newly produced music in terms of listener preference. In addition, as Gioia highlights, at the level of individual songs, streaming rates are also changing in notable ways. In 2021, over 95,000 songs reached 1 million streams. However, in the same year, the 200 most streamed songs counted for slightly less than 1 in 20 of all streams that took place, whereas in 2018, the top 200 songs counted for 1 in 10 streams (ibid.), a point also reflected by Hesmondhalgh et al., who show that the top ten artists accounted for less of than 5 per cent of overall streaming in 2020 (Hesmondhalgh et al., 2021). Clearly, the overall number of streams for music is increasing, and the increase is leaning towards the legacy music of song catalogues. At the same time, as more songs hit larger stream counts, the corollary effect is that the top 200 streaming songs are individually achieving fewer streams. According to Inside Radio, MRC Data showed that in 2019, Lil Nas X's 'Old Town Road' became that year's most streamed song with over 1 billion streams. In contrast, Roddy Ricch's 'The Box' was 2020's most streamed song with 920 million streams, and in 2021, Dua Lipa's 'Levitating' reached the most-streamed position with 627 million streams (Inside Radio, 2022). For Gioia, since old songs are now the cause of growth in the music market, 'the endangered species known as the *working musician* [has] to look on these figures with fear and trembling' (Gioia, 2022). Meier looks at the issue from the perspective of platforms and asks, 'if profit can be accrued from the catalogues of retired and deceased artists, then what reason is there to invest in new music?' (Meier, 2022). This strongly suggests that, as Mercuriadis proposes, music which is familiar, and which may have a variety of emotional and personal resonances for listeners, has the capacity to engage listeners for a variety of reasons, and therefore provide further challenge to musicians wishing to use streaming platforms to directly generate income from their music.

We are in a moment of flux. The meteoric rise of a new asset class in music – song catalogues – and of a global market in music equities, looks set to precipitate significant change in an increasingly financialised digital music sector. Listener trends strongly suggest that the digital music market seems more inclined to reward the already successful, indicating that a divide is emerging between established artists who are able to benefit from listeners' existing connections to their work, and emergent artists who are often challenged by low streaming revenues, as expressed by Nadine Shah in her contribution to the UK Parliament enquiry (Shah, UK Parliament, 2020). While the development and growth of the likes of Hipgnosis may fluctuate, as Meier argues, the interest shown by equity firms in song catalogues 'signifies a reorganisation of productive capacities and responsibilities onto individual artists, record labels and studios' (Meier, 2022). His rationale is that, since it is less risky and less labour-intensive to distribute old music than to create new, then larger companies stand to generate more wealth from legacy music, leaving behind amateur artists, small independent labels and home recording studios, who are very often operating with little-to-no overarching financial security. While his point is a political one, nonetheless, such shifts in security, wealth concentration, listening habits, appetites for innovation and risk management within the music sector are indicative of a new entanglement of productive forces. Measuring music according to an emergent set of technological and financial practices enables us to apprehend a new set of marks that are being left on the body of the practising musician, marks that are reconfiguring musicians' expectations of how they might develop and further themselves and their music within the financialised platform ecology.

Blockchain and Music

In their draft paper, 'How Blockchains Can Support, Complement, or Supplement Intellectual Property', the Coalition of Automated Legal Applications (COALA) set out a compelling vision of a digital economy primed to engage with blockchain, a technology that could dramatically improve opportunities for revenue collection and control of digital content and metadata, for creators and intellectual property rights owners: 'We live in a knowledge economy: a world where value creation is shifting away from moving atoms around, to moving bits around' (De Filippi et al., 2016).

Now halfway through its second decade, blockchain technology is part of a broader paradigm shift in the digital commons, where an array of debates, innovations and movements are seeking to reconfigure economic structures; going beyond the challenge of rethinking business design, to engage more fundamentally with issues of wealth distribution and inclusive governance. Rajan asserts that 'if they are to be free economically', then individuals and small business must be able to own and control their own data (Rajan, 2019: 381). He sees blockchain technology as a key factor in enabling the decentralisation of data ownership, which could in turn lead to a more equitable balance between an economy's three 'pillars': the private sector, the state and the community. Furthermore, given Meier and Manzerolle's (2018) focus on the hybrid nature of music as a digital asset, this suggests that the digital music economy, driven as it is by huge advances in data mining, will find itself increasingly involved with blockchain-based solutions in data and rights management.

Since the mid 2010s, the digital music market has been inundated with a variety of blockchain-based platforms and products, designed to explore and exploit the capabilities of blockchain technologies for engaging with the digital music economy in novel ways. Frequently, an emphasis has been placed on the capacity of blockchain to make the music industry more equitable, both for creators and listeners, by harnessing a variety of technological solutions that are native to blockchain – particularly in relation to protecting creators' rights and piloting a range of payment innovations – along with its underlying governance paradigms of privacy and anonymity. Three notable examples have been the music streaming platforms Resonate, Musicoin and ROCKI, which have explored the affordances of blockchain in terms of business design and the development of tools for creators and listeners, creating platform structures that respond to a set of socio-economic themes; co-operative ownership, a universal basic income (UBI) and the sharing economy.

Resonate: The Ownership of Ownership

'Play fair' (Resonate, 2023a). On its home page, Resonate describes itself as 'the first community-owned music streaming service', announcing in no uncertain terms that it sees itself as a new type of digital music platform. A further click onto the Resonate Blog informs us that 'the mission is to build a new music streaming platform run democratically for the needs and dreams of artists, listeners, and workers (ibid.). From its launch in 2015, Resonate implemented blockchain as both a technological building block and a governance paradigm, operating as a cooperative to explore, at a fundamental level, the implications of fully integrating blockchain into the design and operation of a digital music start-up. In addition to integrating blockchain into its business strategy to develop a new type of platform that was artist centred and would offer 'user-centric stream2own payments' (ibid.), the Resonate ethos clearly has gone beyond a simple interest in using blockchain to improve their performance as a streaming platform, proudly stating that it is 'governed by our artists, members & workers' (ibid.).

As all these excerpts from its website demonstrate, Resonate has had a clear ambition to connect with a variety of narratives that run not only across the music sector, but across a much wider set of contemporary debates, including 'platform cooperativism'.[5] It would seem that the key challenge for Resonate has been to establish a digital music service that could facilitate new online structures for commercial and community association. Although platform cooperativism is a phenomenon that is not necessarily native to blockchain, it is however, an area where blockchain, via its capacity to support immutability, transparency and anonymity, is able to offer a range of solutions for those wishing to build both commercial and not-for-profit projects.

'Nine plays, not 150' (Resonate, 2015). This statement is benchmarked against a calculation that is familiar to many in the music industry, where the sale of one copy of a typical ten-track album is seen to be equivalent to that album being streamed 1,500 times, making the purchase of a single track equivalent to 150 streams (Sisario, 2018). Resonate referenced this nominal figure of 150 streams in its 'Deep dive on stream2own', with the result that artists using its platform stand to receive payment far more rapidly than with conventional approaches to calculating streaming-based payment schedules (Resonate, 2015). The stream2own system scales listener engagement in a way that recognises a change in listening behaviour between casual listeners and those that are more committed. According to this scenario, Resonate's ambition has been to pay artists more quickly than other music streaming platforms (after 9 rather than 150 streams), and with higher payments (ibid.). Clearly users' ownership of music is an important component in the Resonate model, and when it relaunched in 2019, there was a renewed focus on 'scalability and sustainability [and a] fair-trade streaming alternative' (Devlin, 2019). For Resonate founder Peter Harris, the task for the co-operative was not so much about building and running 'a decentralised, distributed database like Bitcoin for metadata and for licensing'; instead he understood that the core challenge was to achieve functional interoperability across data-driven platforms: 'you want to have those qualities and characteristics so that you can show that you're not creating another central data silo, but it doesn't have to go as far as something like Bitcoin' (Harris, 2017). If the ultimate solution to the problems facing a digital music economy is to create an infrastructure and a toolset that can facilitate access and analysis of data across digital platforms and networks, this suggests that what Resonate has been working to resolve is *the ownership of ownership*. Put another way, if, as Meier and Manzerolle show, monetisation of digital commodities and the profitability of music content are increasingly derived from a platform's capacity to own, control and manipulate data, then the Resonate project may yet prove to be an indication of how future innovations in music streaming may involve cooperative ownership, of both platforms themselves as well as digital assets.

Musicoin: Sharism, UBI and Making $MUSIC

Musicoin describes itself as 'a music streaming platform built on the blockchain that supports the creation, distribution and consumption of music in a shared economy' (Musicoin, 2019), or what it refers to as its 'Musiconomy' (Musicoin, 2017). Although its Twitter account has been inactive since March 2022, with the message, 'Dear community members, we are happy to announce that the agreement between Isaac and the new team has been signed!' (Musicoin Project, 2022), the Musicoin website remains live, and its $MUSIC token is still listed on the crypto price-tracking website CoinMarketCap (CoinMarketCap, 2022).

Musicoin uses the term 'musiconomy' to frame the digital music economy as a hybrid matrix of actors that includes miners, songwriters, platform engineers, session musicians, producers, publishers and, indeed, listeners. By bringing together such a disparate mix of stakeholders, traditional music creators, along with worker roles that are native to blockchain and platform ecosystems, Musicoin demonstrates a deep understanding of the way that music production and distribution has been challenged by blockchain and platform technologies. Indeed, it is particularly noteworthy to see an acknowledgement of the value created by listeners, digital music participants who have been the focus of widespread attention, as several blockchain projects have sought to give the role of the listener greater primacy within platform ecosystems. Where Meier and Manzerolle (2018) presented us with the notion of the hybrid music artefact, it may well be that Musicoin's key contribution to the debate about ownership, governance and distribution is this concept of the Musiconomy; where rewards, values, and a set of labours associated with production, mining, curation and sharing are all intermingled to produce a means to interpret and navigate platform economics. Musicoin also operates according to the principle of 'sharism', a concept developed by its founder Isaac Mao (Musicoin, 2019). Sharism manifests as the principles that underpin the Musicoin platform, and is a recognition that digital content contains and creates multiple value streams, and that, by sharing their content, creators are able to generate both monetary and reputational, social value. In the context of the networked, social environment of streaming platforms, Mao's approach is focused on capturing and, to an extent, controlling the way in which music as an intangible asset has an exponential existence both within and beyond of its native platform. In simple terms, the Musicoin operating principle is that although there are clearly revenue opportunities to be had from streaming, the potential to generate indirect rewards from music – both financial and otherwise – are considerable.

Sharism has therefore been a strategy for Musicoin to leverage blockchain technology in a particular way. The platform claims that its 'Pay Per Play (PPP) smart contract is the first of its kind in the cryptocurrency space and is designed exclusively with the interests of musicians in mind' (Musicoin, 2017). The practical consequence of this is that Musicoin not only pays artists for streams that accrue on the platform, but also uses metadata in such a way that enables artists to grow their networks and drive more traffic to their content. The needs of content producers are clearly at the forefront of the platform's design, and creators using Musicoin's services are able to 'retain full ownership of their content' (Musicoin, 2017). The result is that rewards pertaining to intellectual property and the digital music artefact's attendant metadata are all directed towards artists, rather than split with the platform.

Musicoin also makes use of, and adapts, the concept of a Universal Basic Income[6] to support the Musicoin vision of wealth redistribution. The Musicoin whitepaper tells us that 'in Musicoin v2.0, we are introducing a revolutionary new concept in cryptocurrency, "Universal Basic Income (UBI)". UBI is an economic model to ensure each contributor to the platform is fairly rewarded in proportion to their contribution' (Musicoin, 2017). Although there is a certain amount of creative licence at work in Musicoin's use of the UBI concept, nonetheless, there is clearly an attempt to use blockchain to enable a payment system that is not pegged to either a subscription deal for listeners or a reward system that is based on payments to artists that are generated beyond the blockchain.

There are two important issues here. The first is that Musicoin used blockchain to explore business models and emerging governance processes that look beyond current forms. The Musicoin UBI model works to secure an income for musicians from PPP on the platform, at a rate that is not influenced by external market forces (Musicoin, 2019). This suggests that a key focus for the company has been on using blockchain to deliver a payment-no-matter-what principle; an approach that recognises some of the difficulties relating to payments that face musicians that we have seen Resonate engaging with. Second, although the concepts of UBI and sharism are laudable, and suggest an evolved form of ethics for the music industry, Musicoin is actually basing its payment systems on a different form of value creation, one that is native to the cryptocurrency sector. Part of Musicoin's wager for the inherent value of $MUSIC centres on the fact that only a finite amount of the $MUSIC token will ever be produced. In addition, Musicoin operates on the assumption that within a socio-digital network, value can be generated and shared in a variety of ways, including the idea that with increased traffic and platform use, the value of the platform and its token has the potential to increase. As such, provided the underlying cryptocurrency is seen as being a worthwhile investment of its miners' time and effort, then music can be made freely available for listeners to stream on the platform. However, although Musicoin has adopted Bitcoin programmer Satoshi Nakamoto's model of pre-programmed scarcity,[7] this approach may be ill-founded. In her analysis of the problems caused by using objects with inherent value as money, such as precious metals and cryptocurrencies, Mary Mellor argues that 'the original aim [for cryptocurrencies] had been to create a neutral currency to enable trade, bitcoin and similar cryptocurrencies morphed into commodities that were bought as financial investments (Mellor, 2019: 122–3). Mellor's perspective suggests that the value of transactions on platforms like Musicoin (in other words payments made to artists via its UBI protocol) will always be at the mercy of the $MUSIC token's speculative worth as a tradeable commodity. As a result, any artist wishing to use the platform for financial gain is not only putting their faith in the platform's ability to succeed as a business in the competitive and often unrewarding music streaming market, but also in $MUSIC's success as an asset with a fluctuating market value.

Although there are inconsistencies with Musicoin's operating model and recalibration of concepts of sharing and universal basic income, the platform is nonetheless a compelling analogue of the changing face of digital commerce within the blockchain environment. Inherent to Musicoin's design is an acknowledgement that, as platform stakeholders, musicians, miners and listeners are able to contribute to, and benefit from, the platform's success. In this sense, Musicoin can be seen as an attempt to create an ecosystem that harnesses the relationship between token, platform and musical content, and where musicians, listeners and miners brought into a relation that activates social, economic and creative activity around token mining, platform management and practices of making, sharing and listening to music.

ROCKI: 'A Whole New Music Industry and Ecosystem Fit for the Future'

In 2020, Co-founder and CEO of ROCKI, Bjorn Niclas announced the arrival of this new music streaming platform, describing it as 'the first hybrid user-centric blockchain music streaming platform aimed at independent artists and their fans' (Niclas, 2020). The ROCKI whitepaper is bolder in its ambition: 'ROCKI is not trying to fix the music industry of the past, but instead represents a whole new music industry and ecosystem fit for the future' (ROCKI, 2020). Despite the boldness of its claims, the ROCKI project does deliver two key progressions in music streaming: the evolution of blockchain implementation within rights management and payment systems, and the development of the relationship between content creators and listeners. The whitepaper informs us that ROCKI uses 'the strength of the blockchain to handle all aspects where blockchains have proven to be highly effective (transparency, security, trustless contracts, financials)' (ibid.), but unlike other blockchain-powered streaming platforms, it also makes use of fiat, or government-issued, currency, in this case,

US dollars. As such, content creators are paid using a combination of ROCKS cryptocurrency tokens and a fiat-based monthly subscription service that pays creators via a user-centric model, which matches payments to track plays, rather than distributing listener payments across all streams within a given period.

Although user-centric payment models are not native to blockchain systems, it is worth noting that the participants in the UK Parliament enquiry into the economics of music streaming discussed at length the benefits and obstacles of implementing user-centric payment systems within platforms, including the progress made by the French streaming service Deezer (UK Parliament, 2020). Thus, it becomes clear that ROCKI's focus is not simply on building a platform that merely makes use of blockchain in order to overcome certain challenges within the streaming sector. Instead, the approach seems more targeted on identifying the most pressing challenges, and using the most appropriate tool to build solutions. Thus, while the user-centric model purports to reward music creators more effectively than streaming incumbents (such as Spotify), by sharing 'at least 70% of the subscription fee' (ROCKI, 2020), ROCKI uses blockchain to ensure that 'between 20–50 percent of royalty payments' are passed on to artists, by removing the need for intermediaries, such as 'record labels, publishers, distributors, or copyright societies' (ibid.). The platform also uses the language of decentralised ledger technology to describe its artist contracts as 'trustless' (ibid.). It describes its use of smart record contracts as enabling transparency, and a much more dynamic relationship between an artist and their tracks' stakeholders. The whitepaper also suggests that a number of new types of smart contract could also emerge, including 'smart management contracts, smart remix contracts and smart crowdfunding contracts' (ibid.)

The ROCKI whitepaper's description of the platform as an ecosystem, which foregrounds the benefits of the direct artist-to-listener relationship, is also notable, given *Ecologies of Creative Music Practice*'s focus: 'The ROCKI ecosystem [enables artists to] engage with listeners in a bidirectional relationship in stark contrast to the classic "customer / product" relationship that current legacy streaming services offer' (ibid.).

The blockchain-based benefits of 'ecosystem participation' include artists paying listeners their own ROCKS tokens to listen to their music, artists promoting their own tracks via in-site advertising, artists tipping 'superfans' and crowdfunding for up-and-coming artists (ibid.). Such innovations are intended to dramatically improve not only a content creator's experience of music streaming – by adopting a user-centric payment system, increasing the percentage of royalty payments paid to creators, and speeding up payment cycles – but also alter the way in which listeners and artists interact on a streaming platform. By developing tools that enable musicians to incentivise their listeners, this changes not only how listeners might engage with a platform and the music they are consuming, but the dynamic between the platform and an artist also changes. As with Musicoin, ROCKI's design acknowledges the social frame of music, along with a recognition of multiple income streams that can be generated by online music, both direct and indirect, and encourages artists to use financial incentives to create greater listener engagement in order to generate more streams and build their audience base. As such, perhaps the notable contribution to thinking about future iterations of the streaming space that ROCKI poses is its invitation to artists to engage creatively with the platform and to see it as more than simply a shop window that displays their music to potential customers.

Increasingly, ownership in the digital music economy is less a question of owning a musical artefact than a matter of owning the information that surrounds it and, where possible, the mechanism that is used to capture, process and act on that information. Across the three platforms, we have seen different priorities, emphases and innovations around how blockchain technology can rise to this challenge. Where Musicoin developed a language to describe how a blockchain-based platform can facilitate new roles for both listeners and musicians within a blockchain platform ecosystem, Resonate fully embraced not only the capacity, but also the concept, of decentralisation to develop its cooperative platform model. The youngest of the three platforms, ROCKI adopted a hybrid approach to fiat and cryptocurrencies that will facilitate the greatest benefit to all stakeholders within a streaming ecosystem. While the overarching narrative of this chapter acknowledges the ephemeral nature of these and many similar music start-ups, by engaging with these intersections of technology, governance, payment systems and newly calibrated music–listener relations, blockchain-based music systems have nonetheless offered alternative visions of music commerce that harness new forms of user agency and cooperative forms of governance in the hybridised digital music landscape.

Music and NFTs

In 2021, in what was quite possibly one of the many after-effects of the lockdowns caused by the global pandemic, a hitherto largely unheard of blockchain-based token, known as an NFT (a 'non-fungible token') began to gain wider prominence in public awareness. First developed in 2014, NFTs, unlike other blockchain tokens such as Bitcoin and Ether, have a unique identification code (hence their 'non-fungibility'), and therefore can be traded as a digital asset in and of themselves, rather than as a unit of exchange for other goods. NFTs initially gained attention as part of the CryptoKitties game in 2017, where players were able to create and trade NFTs which were now being produced, or 'minted', according to the ERC-721 token standard (Jiang and Liu, 2021). Where the mining process had largely been used to create cryptocurrency tokens and to generate immutable records of transactions, the minting process was now being used increasingly to create unique digital artefacts. These two separate, but related, processes represent the scope of blockchain's capability to interact with music. By aligning music with the mining process, transactions, in the form of proof of ownership, can be registered on a blockchain, and by integrating music with the process of minting NFTs, unique music 'collectibles' – in the form of songs, music-related artwork, videos and the like – can be created.

Through 2021, NFTs and NFT marketplaces proliferated, with the likes of Foundation, KnownOrigin, Nifty Gateway, Rarible, SuperRare and Zora gaining increasing prominence. Notable music-based NFT platforms include Mint Songs, Sound.xyz, Audius, Async Music, Nina, Serenade and Catalog. Serenade makes much of its environmental credentials, foregrounding that it has a smaller carbon footprint than major blockchain players like Ethereum (Mileva, 2022). On its home page, Catalog states that its platform enables users to 'press, trade and listen to one-of-one digital records [that are] artist certified and provably authentic' (Catalog, 2021). The streaming platform ROCKI has also made a foray into NFT trading, with the launch of its own NFT platform (PR Newswire, 2021), and on its homepage it still proudly describes itself as 'The World's Leading Music NFT marketplace' (ROCKI, 2023). As celebrities and major companies engaged with the NFT market in 2021, trading in NFTs reached $17.6 bn, an increase of 21,000 per cent from 2020, and token sales generated profits totalling $5.4 bn (Browne, 2022). According to the cryptocurrency news site CoinDesk, the NFT marketplace OpenSea stood head and shoulders above even its nearest rival, registering over $600 mn in trades in one week alone during October 2021 (Gottsegen, 2021). While the NFT sector has slowed significantly since these peaks, at the time of writing in early 2023, music and media companies' interest in NFTs has persisted. In January 2023, the NFT platform PROOF signed with United Talent Agency (UTA) to expand its market share beyond its core web3 community (Thompson, 2023), and in March 2023 Ticketmaster announced that it had launched a service that will enable artists to reward fans with exclusive access to live event opportunities, such as presales, prime seats, custom travel packages and extra concert experiences in return for purchasing bespoke ticket-NFTs (Ticketmaster Business, 2023). As with all other blockchain tokens, NFT records are stored across a network of computers, and this decentralisation underpins their capacity to remain unique and immutable, given that the network will safeguard an NFT against any attempt to make a copy of it. However, despite being native to the blockchain-based NFT economy, Gottsegen likens OpenSea to Amazon and Microsoft, highlighting the fact that, although it facilitates the trading of digital artefacts that are endowed with unique properties because of their existence on decentralised networks, OpenSea is a private company (ibid.). As a result, there may be a lasting, fundamental and ultimately insurmountable incompatibility between decentralised networks and private corporations that ultimately undermines the NFT ecosystem, given that decentralisation and private ownership tend to pull in opposite directions.

Although the engagement with the NFT market was palpable during 2021, particularly in terms of the excitement generated by concept of one-of-one art and music collectibles, NFTs have had a range of detractors, including those that have highlighted the relationship between NFT minting and carbon emissions (Barber, 2021, Dredge, 2022). David Turner sees that NFTs offer little that is paradigmatically new to the music industry, beyond being another route by which major labels might expand into new markets to offset the diminishing returns of streaming (Turner, 2022). Reflecting more negatively on the impacts that NFTs have had on music, Turner asks who it is that music NFTs are designed to cater for: music fans, and an audience that musicians want to build, or 'people with ETH in their wallet' who have already developed an affinity for blockchain and web3 wallets (Turner, in Baird and Backer, 2022). Turner's lack of enthusiasm for NFTs is appreciable. Essentially, he sees that NFTs can do nothing for music except alienate an artist's traditional fanbase who cannot afford to purchase one-of-one digital collectibles.

Although Turner's ideas are couched within his own broader economic analysis of the music industry, where he sees endemic financialisation and streaming as key challenges to stability for record labels and artists, his views on NFTs resonate with broader perspectives on developments in digital music, where the ongoing development of music as an intangible asset is outpacing traditional models of purchasing and owning music. For example, in a blog post titled 'Intellectual Property Is Becoming Irrelevant', the software engineer Tim Daubenschütz writes that 'for the most part, the internet has stopped caring about intellectual property' (Daubenschütz, 2021). In making this point, Daubenschütz is asserting that what has come to be a conventional approach to engaging with online media is changing. He sees that, increasingly, digital content like music is being understood less as a file or a fixed object, and more like a performance. Daubenschütz's sense is that rights holders are increasingly more concerned with advertising, product placement, building their brand and creating 'performance-based, ephemeral outputs', and that fandom is expressed through a range of engagements with favoured content; shares, likes, mentions and reactions (ibid.). Although Daubenschütz worked at BigchainDB, the blockchain database that underpins Resonate, we should take care not to over-inflate the opinions of a blogger. However, as we shall see in the following chapter, the social-first world of social media seems to be outpacing the content-first world of streaming platforms, through continual creator–consumer engagement and multiple opportunities for monetisation. Although intellectual property clearly does still exist, it is enshrined in a 'pay-per' model of commerce and access to music that itself seems to be evolving, as the socio-technological forces around it shift and change the way that music fans listen to, and interact with, music.

Decentralised Collaboration

Having engaged with examples of specifically music-related platforms and technologies systems, to complete this encounter between music and blockchain-based systems we now turn to two broader conceptions of decentralisation. Doing so will not only enable us to foreground the interdependent nature of creative music practice in terms of its relations with technological environments, but will also serve to remind us of the impact of evolving technological practices on our planetary environment.

In his analysis of Bitcoin, the philosopher Mark Alizart refers to 'Bitcoin [as] a protocol for generating consensus in a decentralised way', suggesting that 'before Bitcoin, it was necessary to choose between consensus and decentralisation' (Alizart, 2020: 28). Alizart's interest in Bitcoin lies in the cryptocurrency's capacity to validate transactions anonymously. Alizart is interested in the way that Bitcoin is composed of three core components: miners, users and coders; the way that these three components interact to create decentralised consensus. In fact, this is the central characteristic from which Alizart draws his view that 'the value of a bitcoin is inseparable from the network that supports it' (ibid.: 34).

Since its inception in 2009, the value of Bitcoin has fluctuated significantly, reaching an initial high of $19k in December 2018, before climbing to a record high in November 2021, when the price of Bitcoin stood at $66.9k. In the year to November 2022, a general decline in the value of Bitcoin reduced its market price to $16.9k (CoinDesk, 2022). Alizart suggests that the problem with Bitcoin is not its volatility, but its lack of liquidity. Alizart is interested in how Bitcoin could be freed from its current cap of 21 million possible coins, and instead, be indexed to global energy consumption (Alizart, 2020: 74). While Alizart's work is not specifically targeted towards analysing the environmental impact of Bitcoin technology, his approach fully acknowledges the direct relations between Bitcoin, energy consumption and the impact on earth systems.[8] In this context, he refers to cryptocurrencies as 'energy currencies', to show that their production and circulation are fundamentally tethered to energy consumption, and sees that cryptocurrencies function 'as a converter between information and energy' (ibid.: 110).

There are two qualities that emerge from Alizart's analysis. The first is that Bitcoin's operating mechanism overlays a form of measurability on energy, meaning that Bitcoin becomes a unit of account for energy flows. Here, Bitcoin itself becomes, in Baradian terms, the boundary produced by practices of measurement and mark making; Bitcoin *is* the measurement of energy and information via technologies of decentralised consensus. Alizart compares the development of Bitcoin to the advent of agriculture and livestock farming, seeing that these innovations reflect humans' capacity to regulate and harness the energy cycles of our environment via 'relations of symbiosis' (ibid.: 110–11). This leads him to describe Bitcoin as 'the currency of life, a "living currency"'(ibid.), where the cryptocurrency becomes the unit of account of our capacity to sustain human culture within our planetary environment, which in itself is a conception of a human–cryptocurrency ecology at a cosmic scale.

Second, Alizart suggests that Bitcoin's capacity to provide decentralised consensus enables it to 'endow information with the quality of energy' (ibid.: 63). In designing Bitcoin, Satoshi Nakamoto's key innovation was to introduce finitude to the practically infinite world of digital information. When energy is moved from one point to another, it does not remain at the first point. It can only be in one place at a time. Information, however, is reproducible, and, when shared, just like a nonrival asset, it can be at one point *and* at a second, third or any number of points at the same time. Alizart's view is that Bitcoin, by limiting the reproducibility of digital information, has created a context where information, like energy, cannot be in two places at once, which, as we have seen, is a cornerstone of the NFT market.

The idea that Bitcoin demonstrates how energy is being replaced by information leads Alizart to propose that cryptocurrencies tend to foster associativity and practices of collaboration. To support this, he builds on the idea that the reproducibility of information has led to humans who share the same language forming 'groups capable of collective intelligence' (ibid.: 87). This again suggests that what we might understand as an ecology of energy and information is precipitating behaviours that emphasise cooperation and mutual assistance, and that cryptocurrency is simply a rendering of this tendency in digital form.

In *Radical Friends: Decentralised Autonomous Organisations and the Arts* (2022), Ruth Catlow and Penny Roberts also address the issue of decentralised consensus, via an extended examination of DAOs.[9] For Catlow and Roberts, DAOs were developed as a means of improving cryptocurrencies through 'coordinating shareholder actions' (Catlow and Roberts, 2022: 28). However, in subsequent years, the capacity of decentralised autonomous organisations to facilitate cooperative rather than hierarchical structures has become more apparent (ibid.: 29). Catlow and Roberts are interested in exploring the way in which DAOs, as collectively owned, member managed, Internet-based organisations that are the product of blockchain development are able to offer new models of membership, coordination and governance (ibid.). Given that DAOs operate according to principles of member ownership and rights, new forms of decentralised consensus have emerged allowing stakeholders to manage assets and services across international borders (Catlow, 2022: 176–7), which offers a longer-term legacy for blockchain technology than either NFTs or mining-based streaming solutions. The writer artist Rhea Myers writes that DAOs use blockchain-based smart contract systems (a means of automating an action once a predefined parameter has been met) 'to structure social forms of human collaboration' (Myers, 2022: 86–7). This is not to say that decentralised systems govern or modify human activity, simply that they offer a new mode for organising and extending collaboration and consensus.

It is notable that the musician Holly Herndon and technology researcher Mat Dryhurst have done much to expand the discourse around use-cases for both DAOs and NFTs in the context of music. In Chapter 5 we shall discuss their Holly+ project in the context of music's encounter with artificial intelligence, but their work is also significant in terms of the current discussion of music rights and income generation. With the meteoric rise of Large Language Model AI systems such as ChatGPT and Google Bard in late 2022 and early 2023, debates over intellectual property rights (IPR) attribution and AI training sets have rapidly gained public attention. In the January 2023 episode of their podcast *Interdependence*, Herndon and Dryhurst discussed the use and attribution of IPR in training sets with the music technologists CJ Carr and Zack

Zukowski, who, as Dadabots, use AI to create music. A Dadabots track featuring Frank Sinatra singing a Britney Spears song that the pair had uploaded to YouTube had recently been taken down by the company because of rights infringement. Given that Herndon and Dryhurst have been exploring rights attribution in the context of AI through a variety of projects, this led to a conversation on the podcast about rights attribution for licensed material that had been in datasets to train AI systems (Interdependence, 2023). As part of Spawning, a group launched in September 2022 consisting of artists, engineers and scientists focused on developing tools that enable creatives to manage their AI identity, Herndon and Dryhurst have worked on two projects that are relevant here.[10] The website *Have I Been Trained* enables artists to see when their work has been used in training sets for image creation systems such as Open AI's DALL-E and Stable Diffusion (Spawning, 2023), while *Source+* allows artists to opt in to or out of these training sets (Stokel-Walker, 2022).

The Holly+ project makes use of a DAO network in combination with NFTs as a means to acknowledge and remunerate contributors to music made using the Holly+ protocol. On the Holly+ website, Herndon (2021) describes how NFTs are being used:

- Artist X produces a song using the Holly+ voice model.
- Artist X uploads the song online, and submits the song as a proposal to the DAO through a public interface.
- VOICE token holders vote to mint the song as an appropriate or inspiring usage of the Holly+ voice.
- An NFT of the song is minted by the DAO, with 50% of sales generated going to the artist, and 40% of sales going to DAO members, and 10% reserved for me.

Herndon details how stakeholders in the Holly+ DAO act as 'stewards' of the voice model, and are incentivised to certify or license new works that add value to the project, and to Herndon's voice, rather than diluting its value by producing 'bad art or negative associations' (ibid.). Thus, we see a rights trajectory that uses NFTs to incorporate and acknowledge the work of different types of contributors – creators and copyright holders within training sets – while simultaneously using decentralised governance paradigms which are designed to emphasise aesthetic quality and reputational benefit, so as to curate new work within the Holly+ brand.

Such emergent theories of consensus, and experiments in using blockchain technology to propagate new forms of creative collaboration and remuneration, offer valuable perspectives on the ecological nature of music, and music practices. Catlow and Roberts' (2022) interest in the inherently collaborative and associative dynamics of DAOs emphasises the cooperative trajectories pursued by the likes of Resonate, suggesting that blockchain technologies still have much to offer musicians beyond simply operating as an enhanced auditing system within current industry paradigms. DAOs as a form of platform cooperativism, wherein artists and musicians go beyond simply collecting payments from listeners making use of their intellectual property, and instead participate in collective ownership of music platforms, suggest a considerable change for musicians; from owners of rights, to co-owners of systems and participants in governance frameworks and stakeholder networks. Alizart looks to Bitcoin's underlying dependence on networks to assert that Bitcoin is nothing if not an embodiment of consensus. At the same time, his perspectives on how Bitcoin has catalysed a reciprocity between information and energy provides a more holistic vision of the necessary interrelations between information, energy and collective behaviours. The consequences of this are twofold. First, he contends that, through Bitcoin, information has taken on the non-reproducible qualities of energy. As a living currency that is benchmarked against global energy use, Alizart suggests that in the form of Bitcoin, information can be increasingly accounted for both in terms of its relation to earth systems and to human systems that are built around networks of shared information. As we have seen, music platforms and economies built on blockchain systems are exactly these types of networks. Second, given that Bitcoin bears the marks of information's capacity for limitless reproducibility, then in the same way that information's reproducibility has facilitated human development through acts of collective intelligence, blockchain technology will continue to propagate collaboration and consensus. In the context of the blockchain paradigm, particularly issues of digital information, consensus and energy accounting, we can also consider music in terms of these qualities, where music-as-information becomes a means to facilitate consensus and cooperation, and to measure our impact on our planetary environment.

Conclusions: Entangled Creativity

There is much to be learned from these experiments with catalogue ownership, music streaming and blockchain technology. Fundamentally, this chapter has examined how music's recent evolution into an intangible and hybrid digital asset has begun to fundamentally transform how music is owned, and the consequences of this transformation for both musicians and listeners. These are matters of access; both to music itself, and to the opportunity to earn an income from music.

Ownership is changing in multiple and significant ways; ownership of products, of intellectual property, and opportunities to own a stake in the platform infrastructures that surround, contain and enable these evolving forms. Blockchain technology has clearly played a noteworthy role in how a variety of services might be imagined within the music streaming sector, and stands to continue offering new opportunities and challenges to established notions of what constitutes a music creator and a music consumer. In a more general sense, we have also seen that the evolution of music into an intangible hybrid of rights and data, which is being used in myriad ways by a variety of platforms and funds, shows little sign of slowing down.

Where Garvey et al. have highlighted the challenges facing new and future artists which result from the infrastructure of the music streaming economy (UK Parliament, 2020), the likes of Hipgnosis Songs Fund are shoring up the fortunes of artists who have already amassed a significant market share for their music, with a highly reliable and multi-faceted listener base. On this model, it would seem that legacy artists have much to gain from equities markets. However, beyond the likes of superstar musicians such as Justin Bieber, for many artists working to establish and grow a financially viable career, obstacles such as low royalty returns on streaming platforms mean that the route to becoming a legacy artist looks set to become increasingly challenging. Given this increasing divergence in the music sector between legacy artists and emerging artists, it is notable that while the former are benefitting from the emergence of songs and music funds trading on equities markets, blockchain platforms have focused their attention on the latter. ROCKI's ambition to build an entirely new music industry is notable in this regard, with their stated aim to 'avoid legacy record labels and institutions' (ROCKI, 2020). However, although the platform's confidence in the ongoing growth of independent sector is compelling, this is in stark contrast to the acquisitions and funding rounds that characterise the catalogue sector. This difference also speaks of the relative challenges and opportunities associated with creating new music and generating revenue from existing music. It is clear that catalogue funds are benefitting from public appetites for established and familiar styles, which means that they need not concern themselves with the risks associated with discovering, supporting and funding new artists, and the expense involved in developing and promoting them. This risk and expense, identified by Tom Gray and Nadine Shah in the UK government enquiry has not gone away, it has simply been transferred to the emergent artists themselves (UK Parliament, 2020). If streaming royalties are not sufficient to compensate early-career musicians, then the likes of ROCKI are launching and evolving to support them, although as we shall see in the next chapter, the social music economy also offers opportunities and challenges to the conventional model of royalty payments for audio-only musical content.

Isaac Mao's concept of a 'Musiconomy' frames how streaming platforms and, in a wider sense, the music economy as a production infrastructure, can be efficiently designed and produced via a participatory, bottom-up process. Similarly, the Resonate manifesto informs us that 'co-ops are the future of an equitable internet. Technology should benefit all involved, not just a handful of shareholders' (Resonate, 2023b). Resonate's strategy to increase cooperative ownership of data within its platform demonstrated a profound commitment to the notion that greater involvement can produce greater autonomy with regard to strategising the use of metadata, a fundamental principle of providing access to and control of data users that we see being further explored by the likes of Herndon and Dryhurst across their various projects.

The musiconomy model is an ecological approach to thinking about contemporary music infrastructures; it is an acknowledgement of the interdependent nature of the different participants and components within a platform structure: the technological system itself, the technologists who maintain the system, the commercial and governance frameworks, the music creators and the listeners, all operating in relation to each other, and creating different types of value for different aspects of a platform. While still marginal, the blockchain streaming sector offers multiple perspectives on possible alternatives and modifications to the digital music sector. Beyond a standard blockchain model where musicians create content, miners create tokens, and listeners create streams, in a blockchain ecology musicians can also create payments, listeners can also create promotional materials, and miners can also create value by buying and selling tokens on currency exchanges. Again, the Resonate and ROCKI models demonstrate how this is quite literally being folded into an emergent model of ownership, whereby platform users and owners can access and manipulate their metadata (in the case of ROCKI), and co-owners have been able to control their data and gain reward as the platform grows (Resonate). In this sense, ownership is part of a socio-digital contract with a platform, that sets not only the specific rules and terms of engagement for platform use, but also defines a context within which platform users and contributors can expect to participate. Moreover, in cases where a blockchain platform has committed to giving musicians greater control of their own data, as in the case of Holly+, such agreements are an acknowledgement of mutual benefit, and designed to secure long-term futures for each party.

As with Herndon and Dryhurst's commitment to propagating the use of NFTs in practices of mutual association and cooperation in music production, Catlow and Roberts' (2022) and Alizart's (2020) technological, creative, political and philosophical perspectives, while not specifically focused on music, also encourage us to think beyond blockchain-based music streaming and NFT music collectibles. Following Alizart's proposals for living and energy currencies, we can see music as a unit of account for energy consumption and management, converting music's shareability and abundance into an energy-like, non-reproducible form. Similarly, in terms of notions of consensus and collective action, the streaming innovations of Musicoin, Resonate and ROCKI have all attempted to address associativity in their own ways, albeit within the context of a strictly commercialised environment. Connecting creators to platform developers, miners and listeners, creating incentivisation opportunities for music fans, and building contexts for cooperative action by protecting individual rights are all ultimately experiments in decentralised governance; explorations of music, connectedness and commerce within a blockchain-inflected ecology.

The intangible and hybrid nature of music as a digital asset is generating a radically new economic underpinning for how a range of stakeholders are engaging with music. At this point, it is far from clear what the longer-term impacts of the most recent phase of financialisation in music will be, and how the emergence of catalogue funds will impact the wider music sector. Neither can we be certain what role blockchain will ultimately play in shoring up the future of music making, or whether it will foster an age of mutual association and collaboration. While the developments in blockchain-based music projects have been dynamic, and have enabled pioneers, technologists, entrepreneurs and musicians to accumulate significant financial reward, the technological innovations themselves have tended to be relatively short-lived. In ecological terms, these evolutions in music are an inherent consequence of the interdependent relations between music, the technological and financial structures that support its distribution and consumption, and of our planet's energy systems.

Interdependence in music manifests as a perpetual state of change, and blockchain technology, by considerably extending the hybridity of music as an intangible asset, has offered new perspectives on how music can be monetised and valued. On platforms where the previously fixed roles of creator and consumer have become malleable, the use of blockchain has shown that these roles are not permanent categories. Instead, we must understand them in relation their place within a given system. As we have seen in previous chapters through the work of Barad, Laruelle, Meillassoux and Morton, change is palpable and not simply a product of our capacity to experience it. This means that the fluctuations in music practice and music commerce really are transforming production and consumption habits, and having tangible impacts on traditional conceptions of making, sharing and listening to music. Barad's theories enable us to understand how, in an ongoing process of entanglement, musicians, music fans, listeners, platform owners and rightsholders will all engage with practices of making, sharing and listening to music, but importantly, none of these are fixed properties. Each entangled strand does not pre-exist or survive the entanglement, and each results from its intra-action with the strands around it. Musicians form, and are formed by, their intra-actions with blockchain platforms and catalogue funds. Indeed, as a consequence of its interrelatedness with technology and commerce, the seemingly fundamental category 'musician' is being reformed as technological paradigms and commercial practices alter around it. Musicians are no longer simply people who make music. Musicians are becoming AI trainers, blockchain producers, asset traders and platform owners. What is more, by creating music in the form of intangible assets, musicians are not simply producing music; they are producing digital units of account for planetary energy consumption. These are real changes that are transforming working practices, not simply transient perspectives reflecting on technological trends. Where measurements are boundary-forming and entanglements leave marks on bodies, the boundaries and marks we have encountered in this chapter include the platforms, the NFTs, the song catalogues and the legacy artists that populate the intangible music economy. All are momentary points in the ongoing evolution of music, and, as we progress into an encounter with the Creator Economy in the next chapter, we shall see further evidence of change and evolution in creative practices of music.

Notes

1 Spotify describes its Loud&Clear data portal as a means of 'sharing new data on Spotify's royalty payments and breaking down the global streaming economy, the players, and the process' (Loud&Clear, 2022).

2 Here we can see Rajan drawing on Edward Freeman's 'stakeholder' theory. According to Stieb (2009), Freeman's ideas are focused on reconfiguring traditional approaches to business, such that benefits and 'important decision-making powers' should be redistributed beyond 'stockholders', to a wider network of 'stakeholders'. Stieb quotes Freeman: 'The crux of my argument, is that we can reconceptualize the firm around the following question. For whose benefit and at whose expense should the firm be managed?' (Freeman, in Stieb, 2009: 405). Freeman's project was concerned with re-evaluating theories of the firm, and in particular, challenging the accepted framework of managerial capitalism, by recognising the fundamental importance of a firm's interaction with a network of stakeholders. Elsewhere, Freeman positions stakeholder theory as a means of understanding 'value-creation activity as a contractual process among those parties affected; [designating] those parties as financiers, customers, suppliers, employees, and communities' (Freeman, 1994: 415). Freeman was interested in moving beyond traditional approaches to understanding the purpose of businesses, and he challenged the accepted view that firms are fundamentally designed to create value for shareholders. His alternative was to recognise a wide set of interests in the affairs of a given firm, along with a heterogeneous set of values that would be generated in relation to these different interest groups. Stieb quotes Freeman's call for a given firm 'to interact with other communities that it affects or is affected by, seeking to understand their perspectives, listen to their preferences, and evaluate the impact of actions on them'. Freeman's ecological approach was to frame this interaction between stakeholders as 'cooperation with those upon whom it relies for support – employees, suppliers and customers' (Freeman, in Stieb, 2009: 405).

3 The online royalty auction platform and marketplace, Royalty Exchange introduced the Dollar Age metric as a means to predict the future performance of a song in relation to its age and its earnings within the previous 12 months. The company bases its approach on the idea that 'the longer something has been around, the longer it is likely to continue to stay around' and therefore proposes that Dollar Age 'allows investors to better assess the potential risk of unfamiliar catalogues that may still hold value' (Royalty Exchange, 2021).

4 MRC Data, which Gioia and others refer to in their own reports, was renamed 'Luminate' in March 2022 (Spangler, 2022).

5 The term 'platform cooperativism' was first used by the academic Trebor Scholz in 2014, in an article entitled 'Platform Cooperativism vs. the Sharing Economy' (Scholz, 2014). Scholz clearly intended to cause a shift in public understanding and attitudes towards platform disruptors such as Airbnb, Uber and TaskRabbit, who, in the early 2010s, were very much in the ascendant. These 'lean platforms' (Srnicek, 2017) were building new business frameworks and providing seemingly radical alternatives for huge numbers of people to create revenue from their hitherto under- or unexploited skills, resources and time. Also writing in 2014, Nathan Schneider took a similarly dim view of the sharing economy, writing in 'Owning Is the New Sharing' (Schneider, 2014) that, although the lean platforms were seemingly creating a future economy where ownership could be replaced by a culture of renting (to which the euphemism 'sharing' neatly provided a networked, utopian sheen), it was 'becoming clear that ownership matters as much as ever ... whoever owns the platforms that help us share decides who accumulates wealth from them' (Schneider, 2014). Although Schneider's and Scholz's work can now be seen as part of a broader context of discourse that is currently focused on the economic imbalances and asymmetries in the sharing platform model, the concept of platform cooperativism has clearly played a significant role in creating credible alternatives to the standard

sharing economy model. In this sense, it is worth considering the breadth of Scholz's vision for 'worker–owned cooperatives' and worker-designed 'apps-based platforms': '*platform cooperativism* can invigorate genuine sharing, and it does not have to reject the market ... it can be a reminder that work can be dignified rather than diminishing for the human experience [and] could help to weave some ethical threads into the fabric of 21st century work' (Scholz, 2014).

6 Musicoin's tethering of fair reward to a proportion of an artist's creative contribution diverges significantly from a more standard definition of UBI. For example, the *Royal Society for the Encouragement of Arts, Manufactures and Commerce* frame a UBI as something that provides 'every individual with a foundation of a regular, unconditional, cash payment' (Painter, 2019). Similarly, a preliminary report for the Finnish Government, produced as a precursor to the experimental roll-out of a Universal Basic Income during 2017–2018, sets out a UBI as a 'basic income [that] is paid to all individuals aged 18 and over but not to pensioners (old-age pensions, disability pensions)' (Kirjoittaja ja Kela, 2015: 43).

7 Satoshi Nakamoto programmed Bitcoin in such a way to ensure that a maximum of 21 million coins could be mined. In an email exchange dating from 2009 with fellow Bitcoin developer, Mike Hearn, Nakamoto reasoned that 'there's only going to be 21 million coins for the whole world, so it would be worth much more per unit' (Nakamoto, 2009).

8 The debate around Bitcoin's carbon footprint is well established. Writing in the journal *Scientific Reports*, Jones, Goodkind and Berrens show that Bitcoin mining can be compared to 'energy intensive or heavily-polluting commodities such as beef production, natural gas electricity generation, or gasoline from crude oil' (Jones, Goodkind and Berrens, 2022).

9 Decentralised Autonomous Organisations, or DAOs, initially gained wider attention in 2016 when a DAO launched by the start-up company Slock.it – known simply as 'The DAO' – was hacked, leading to the loss of 3.6 million Ether tokens, then valued at $50 mn (Seigel, 2022).

10 The group use the term 'spawning' to refer to the process of using AI systems to create new media, by training those systems on existing media. The term is designed to differentiate the process from sampling and collage, since AI systems do not re-present existing material within new works, instead they learn stylistic traits and approaches from existing datasets (Spawning, 2023).

Bibliography

Alizart, M. 2020. *Cryptocommunism*. Cambridge: Polity Press.

Allen, D., Berg, A. and Markey-Towler, B. 2019. 'Blockchain and supply chains: form organisations, value redistributions, de-commoditisation and quality proxies'. *The Journal of The British Blockchain Association*, 1, pp. 1–8.

Aswad, J. 2023. 'Justin Bieber sells music rights to Hipgnosis songs for $200 million-plus'. Available at https://variety.com/2023/music/news/justin-bieber-sells-music-rights-hipgnosis-200-million-1235497225/ (accessed April 2023).

Baird, S. and Backer, S. 2022. 'Web 3.Bro with David Turner'. Available at https://www.podbean.com/ew/pb-np6xd-11a0004 (accessed November 2022).

Barabási, A-L. 2016. *Network Science*. Cambridge: Cambridge University Press. Available at http://networksciencebook.com (accessed November 2022).

Barber, G. 2021. 'NFTs are hot. So is their effect on the Earth's climate'. Available at https://www.wired.com/story/nfts-hot-effect-earth-climate/ (accessed November 2022).

Browne, R. 2022. 'Trading in NFTs spiked 21,000% to more than $17 billion in 2021, report says'. Available at https://www.cnbc.com/2022/03/10/trading-in-nfts-spiked-21000percent-to-top-17-billion-in-2021-report.html (accessed November 2022).

Catalog. 2021. 'Catalog: Where music is worth it'. Available at https://catalog.mirror.xyz/-1y7KKNNO6V4HIHfZEvo1M9BJ2jAL4oDkPx7sCAw0xM (accessed November 2022).

Catlow, R. 2022. 'Translocal belonging and cultural cooperation after the blockchain – A citizen sci-fi', in Catlow, R. and Roberts, P. (eds), *Radical Friends: Decentralised Autonomous Organisations and the Arts*. Torque Editions.

Catlow, R. and Roberts, P. 2022. *Radical Friends: Decentralised Autonomous Organisations and the Arts*. Torque Editions.

Clark, B and Burstall, R. (2019). 'Crypto-pie in the sky? How blockchain technology is impacting intellectual property law'. *Stanford Journal of Blockchain Law & Policy*. Available at https://stanford-jblp.pubpub.org/pub/blockchain-and-ip-law (accessed November 2022).

Clayton, R. and Clayton, C. 2022. 'UK screen use in 2022: A need for guidance'. Available at https://www.leeds.ac.uk/policy-leeds/doc/uk-screen-use-2022 (accessed November 2022).

CoinDesk. 2022. 'Bitcoin'. Available at https://www.coindesk.com/price/bitcoin/ (accessed November 2022).

CoinMarketCap. 2022. 'Musicoin'. Available at https://coinmarketcap.com/currencies/musicoin/ (accessed November 2022).

Daubenschütz, T. 2021. 'Intellectual property is becoming irrelevant'. Available at https://timdaub.github.io/2021/08/22/intellectual-property-will-become-irrelevant/ (accessed November 2022).

De Filippi, P., McMullen, G., McConaghy, T., Choi, C., de la Rouviere, S.Benet, J. and Stern, D. J. 2016. *How Blockchains Can Support, Complement, or Supplement Intellectual Property: Working Draft, Version 1.0*. COALA IP. Available at https://www.intgovforum.org/en/filedepot_download/4307/529 (accessed November 2022).

Devlin, K. 2019. 'Relaunched music sharing platform, resonate, operates "stream to own model"'. Available at https://djmag.com/news/relaunched-music-sharing-platform-'resonate'-operates-"stream-own"-model (accessed November 2022).

Dredge, S. 2020. 'Spotify CEO talks Covid-19, artist incomes and podcasting'. Available at https://musically.com/2020/07/30/spotify-ceo-talks-covid-19-artist-incomes-and-podcasting-interview/ (accessed November 2022).

Dredge, S. 2022. 'Are "eco-friendly" NFTs as green as their creators claim?' Available at https://musically.com/2022/02/10/are-eco-friendly-nfts-green/ (accessed November 2022).

Freeman, E.R. 1994. 'The politics of stakeholder theory: Some future directions'. *Business Ethics Quarterly*, 4(4), pp. 409–421.

Gioia, T. 2022. 'Is old music killing new music?' Available at https://tedgioia.substack.com/p/is-old-music-killing-new-music (accessed November 2022).

Gottsegen, W. 2021. 'The NFT market is already centralized'. Available at https://www.coindesk.com/tech/2021/10/18/the-nft-market-is-already-centralized/ (accessed November 2022).

Harris, P. 2017. 'Blockchain music co-op challenges Spotify – featuring Peter Harris'. Available at https://blockchain.global/peter-harris/ (accessed November 2022).

Haskel, J. and Westlake, S. 2018. *Capitalism Without Capital: The Rise of the Intangible Economy*. Princeton and Oxford: Princeton University Press.

Herndon, H. 2021. 'Holly+'. Available at https://holly.mirror.xyz/54ds2IiOnvthjGFkokFCoaI4EabytH9xjAYy1irHy94 (accessed May 2022).

Hesmondhalgh, D., Osborne, R., Sun, H. and Barr, K. 2021. *Music Creators' Earnings in the Digital Era*. Available at https://www.gov.uk/government/publications/music-creators-earnings-in-the-digital-era (accessed October 2022).

Hipgnosis. 2021. 'Hipgnosis'. Available at https://www.hipgnosissongs.com/home (accessed November 2022).
Hipgnosis Songs Fund Limited. 2020. 'Interim report 2020'. Available at https://www.hipgnosissongs.com/home (accessed November 2022).
Hu, C. 2020. 'Water and music. "Hipgnosis songs fund"'. Available at https://mailchi.mp/cheriehu/hipgnosis?e=15f0232717 (accessed November 2022).
Ingham, T. 2020a. 'Round Hill fund raises $282m ahead of going public this Friday'. Available at https://www.musicbusinessworldwide.com/round-hill-fund-raises-282m-ahead-of-london-ipo-this-friday/ (accessed November 2022).
Ingham, T. 2020b. 'I was shocked the major music companies even let us in the business'. Available at https://www.musicbusinessworldwide.com/podcast/larry-mestel-i-was-shocked-the-majors-even-let-me-in-the-business-back-in-2006/ (accessed November 2022).
Ingham, T. 2021. 'Hipgnosis acquires 50% of Neil Young's song catalog for around $150m'. Available at https://www.musicbusinessworldwide.com/hipgnosis-acquires-50-of-neil-youngs-song-catalog-for-around-150m/ (accessed November 2022).
Inside Radio. 2022. 'MRC data: Top hits have less reach as "streaming's torso grows"'. Available at https://www.insideradio.com/free/mrc-data-top-hits-have-less-reach-as-streaming-s-torso-grows/article_97f6ea98-7378-11ec-9bc6-533a1e56c0fd.html (accessed November 2022).
Interdependence. 2023. 'Dadabots: Playing with fire?and extreme AI music'. Available at https://interdependence.fm/episodes/dadabots-playing-with-fireand-extreme-ai-music (accessed April 2023).
Kirjoittaja ja Kela. 2015. 'From idea to experiment: Report on Universal Basic Income Experiment in Finland'. Available at https://basicincome.org/wp-content/uploads/2016/07/KELA_Preliminary_Report_UBI_Pilots.pdf (accessed November 2022).
Jiang, X.-J., and Liu, X.F. 2021. 'CryptoKitties transaction network analysis: The rise and fall of the first blockchain game mania'. *Frontiers of Physics*, 9, doi:10.3389/fphy.2021.631665
Jones, B.A., Goodkind, A.L. and Berrens, R.P. 2022. 'Economic estimation of Bitcoin mining's climate damages demonstrates closer resemblance to digital crude than digital gold'. *Scientific Reports*, 12, 14512. doi:10.1038/s41598-022-18686-8
JP Morgan. 2019. 'The next step for blockchain: What are the barriers and opportunities for the evolving blockchain ecosystem?' Available at https://www.jpmorgan.com/global/research/blockchain-next-steps (accessed November 2022).
Loud&Clear. 2022. 'About'. Available at https://loudandclear.byspotify.com (accessed November 2022).
Mao, I. 2010. 'Isaac Mao on the concept of sharism'. Available at https://www.youtube.com/watch?v=jFBMgknTK8M (accessed November 2022).
Malherbe, L., Montalban, M., Bédu, N. and Granier, C. 2019. 'Cryptocurrencies and blockchain: Opportunities and limits of a new monetary regime'. *International Journal of Political Economy*, 48 (2), pp. 127–152. doi:10.1080/08911916.2019.1624320
Mazzucato, M. 2013. 'Taxpayers helped Apple, but Apple won't help them'. Available at https://hbr.org/2013/03/taxpayers-helped-apple-but-app (accessed August 2022).
Mazzucato, M. 2018. *The Entrepreneurial State: Debunking Public vs. Private Sector Myths*. London: Penguin Books.
Meier, G. 2022. 'Sound money: On bandcamp, neil young derivatives and the financial imagination of music production'. Available at https://bellonamag.com/sound-money (accessed November 2022).

Meier, J. and Manzerolle, V. 2018. 'Rising tides? Data capture, platform accumulation, and new monopolies in the digital music economy'. *New Media & Society*, 21(3), pp. 543–561.

Mellor, M. 2019. *Money: Myths, Truths and Alternatives*. Bristol, Chicago: Policy Press.

Mercuriadis, M. 2021. 'Merck Mercuriadis, CEO of Hipgnosis talks about music royalties as an asset class'. Available at https://www.youtube.com/watch?v=AtGcKyy7OBU (accessed November 2022).

Mileva, G. 2022. 'The top 25 music NFT marketplaces to find and sell music NFTs'. Available at https://influencermarketinghub.com/music-nft-marketplace/ (accessed November 2022).

Music Business Worldwide. 2021a. 'Hipgnosis songs fund'. Available at https://www.musicbusinessworldwide.com/companies/hipgnosis-songs-fund/ (accessed November 2022).

Music Business Worldwide. 2021b. 'Round Hill Music'. Available at https://www.musicbusinessworldwide.com/companies/round-hill-music/ (accessed November 2022).

MusicAlly. 2020. 'Primary Wave buys stake in the royalties of artist Air Supply'. Available at https://musically.com/2020/02/20/primary-wave-buys-stake-in-the-royalties-of-artist-air-supply/ (accessed November 2022).

Musicoin. 2017. *Musicoin: A decentralized platform revolutionizing creation, distribution and consumption of music. Whitepaper Version 2.0.0*. Available at https://drive.google.com/file/d/1KVvcwPKUngMNffgWW65k1p4UvKg5QG0u/view (accessed November 2022).

Musicoin. 2019. 'How it works'. Available at https://musicoin.org/how-it-works (accessed January 2021).

Musicoin Project. 2022. [Twitter]. 11 March. Available at https://twitter.com/musicoins/status/1502363114892939269 (accessed November 2022).

Myers, R. 2022. 'A thousand DAOs', in Catlow, R. and Roberts, P. (eds), *Radical Friends: Decentralised Autonomous Organisations and the Arts*. Torque Editions.

Nakamoto, S. 2009. 'Mike Hearn emails'. Available at https://www.bitcoin.com/satoshi-archive/emails/mike-hearn/#selection-13.1-13.18 (accessed November 2022).

Niclas, B. 2020. 'Introducing ROCKI'. Available at https://medium.com/rockiapp/introducing-rocki-d628ad2d242e (accessed November 2022).

Ostrom, E. 2010. 'Beyond markets and states: Polycentric governance of complex economic systems'. *American Economic Review*, 100(3), pp. 641–672. doi:10.1257/aer.100.3.641

Painter, A. 2019. 'The case for basic income is growing. Scotland can take it forward'. Available at https://www.thersa.org/discover/publications-and-articles/rsa-blogs/2019/05/ubi-scotland (accessed November 2022).

Panisi, F., Buckley, R.P., and Arner, D. 2019. 'Blockchain and public companies: A revolution in share ownership transparency, proxy voting and corporate governance?' *Stanford Journal of Blockchain Law & Policy*. Available at https://stanford-jblp.pubpub.org/pub/blockchain-and-public-companies (accessed November 2022).

Pejic, I. 2019. *Blockchain Babel: The Crypto Craze and the Challenge to Business*. London, New York, New Delhi: Kogan Page.

Platform Cooperativism Consortium. 2019. 'Vision & advantages'. Available at https://platform.coop/about/vision-and-advantages/ (accessed November 2022).

Powell, J. 2021. 'Stifel is worried that Hipgnosis Songs Fund is slipping out of tune'. Available at https://www.ft.com/content/30051971-9b61-4b2b-ba9e-64a2c0df0378 (accessed November 2022).

PR Newswire. 2021. 'ROCKI launches the world's biggest Music NFT platform dedicated to musicians and fans on Binance Smart Chain (BSC)'. Available at https://www.prnewswire.com/news-releases/rocki-launches-the-worlds-biggest-music-nft-pla

tform-dedicated-to-musicians-and-fans-on-binance-smart-chain-bsc-301282968.html (accessed November 2022).

Primary Wave. 2020. 'Primary Wave partnership with the award-winning and multi-platinum selling duo Air Supply'. Available at https://primarywave.com/primary-wave-music-publishing-announces-they-have-acquired-a-stake-in-the-music-publishing-catalog-and-artist-master-royalty-stream-of-the-award-winning-and-multi-platinum-selling-duo-air-supply/ (accessed November 2022).

PwC. 2019. 'Blockchain is here. What's your next move?' Available at https://www.pwc.com/gx/en/issues/blockchain/blockchain-in-business.html (accessed November 2022).

Rajan, R.G. 2019. *The Third Pillar: How Markets and the State Leave the Community Behind*. India: HarperCollins.

Resonate. 2015. 'Deep dive on stream2own'. Available at https://resonate.is/stream2own-explained/ (accessed November 2022).

Resonate. 2023a. 'Resonate'. Available at https://resonate.is (accessed March 2023).

Resonate. 2023b. 'It's time to play fair'. Available at https://resonate.is/manifesto/ (accessed March 2023).

ROCKI. 2020. 'ROCKI: A hybrid user-centric music streaming ecosystem powered by the rocks token promotions'. Available at https://rocki.com/white-paper.pdf (accessed April 2023).

ROCKI. 2023. 'ROCKI: The world's leading NFT marketplace'. Available at https://rocki.com (accessed April 2023).

Round Hill Music Royalty Fund Limited. 2021. 'Music royalties with a proven track record'. Available at https://www.roundhillmusicroyaltyfund.com (accessed November 2022).

Royalty Exchange. 2021. 'What is Dollar Age? Introducing Dollar Age: A standardized metric to evaluate royalty revenue stability'. Available at https://www.royaltyexchange.com/blog/what-is-dollar-age (accessed November 2022).

Sanchez, D. 2017. 'What streaming music services pay (updated for 2017)'. Available at https://www.digitalmusicnews.com/2017/07/24/what-streaming-music-services-pay-updated-for-2017 (accessed November 2022).

Sanchez, D. 2018. 'What streaming music services pay (updated for 2022)'. Available at https://www.digitalmusicnews.com/2018/12/25/streaming-music-services-pay-2019/ (accessed November 2022).

Schneider, N. 2014. 'Owning is the new sharing'. Available at https://www.shareable.net/owning-is-the-new-sharing/ (accessed November 2022).

Scholz, T. 2014. 'Platform cooperativism vs. the sharing economy'. Available at https://medium.com/@trebors/platform-cooperativism-vs-the-sharing-economy-2ea737f1b5ad (accessed November 2022).

Seigel, D. 2022. 'Understanding The DAO attack'. Available at https://www.coindesk.com/learn/2016/06/25/understanding-the-dao-attack/ (accessed November 2022).

SESAC. 2022. 'About SESAC'. Available at https://www.sesac.com/our-history (accessed November 2022).

Sherwin, A. 2014. 'Free U2 album: How the most generous giveaway in music history turned PR disaster'. Available at https://www.independent.co.uk/arts-entertainment/music/features/free-u2-album-how-the-most-generous-giveaway-in-music-history-turned-into-a-pr-disaster-9745028.html (accessed November 2022).

Sisario, B. 2018. 'The music industry's math changes, but the outcome doesn't: Drake is no. 1'. Available at https://www.nytimes.com/2018/07/09/arts/music/drake-scorpion-streams-billboard-chart.html (accessed November 2022).

Spangler, T. 2022. 'Luminate is the new name of P-MRC data, source of music and entertainment industry data'. Available at https://variety.com/2022/digital/news/luminate-p-mrc-data-music-entertainment-1235205741/ (accessed November 2022).

Spawning. 2023. 'AI tools for artists. Made by artists'. Available at https://spawning.ai (accessed April 2023).

Srnicek, N. 2017. *Platform Capitalism*. Cambridge: Polity Press.

Stassen, M. 2022. 'Primary Wave strikes $2 billion deal with financial giant Brookfield to acquire music copyrights'. Available at https://www.musicbusinessworldwide.com/primary-wave-strikes-2-billion-deal-with-financial-giant-brookfield-to-acquire-music-copyrights/ (accessed November 2022).

Stassen, M. 2023. 'Done Deal: Justin Bieber sells catalog to Hipgnosis' Blackstone fund'. Available at https://www.musicbusinessworldwide.com/justin-bieber-sells-catalog-hipgnosis-blackstone-fund/ (accessed April 2023).

Steele, A. 2020. 'Stevie Nicks sells stake in songwriting catalog'. Available at https://www.wsj.com/articles/stevie-nicks-sells-stake-in-songwriting-catalog-11607095635 (accessed November 2022).

Stieb, J.A. 2009. 'Assessing Freeman's stakeholder theory', *Journal of Business Ethics*, 87 (3), pp. 401–414. doi:10.1007/s10551-008-9928-4

Stokel-Walker, C. 2022. 'This couple is launching an organization to protect artists in the AI era'. Available at https://www.inverse.com/input/culture/mat-dryhurst-holly-herndon-artists-ai-spawning-source-dall-e-midjourney (accessed April 2023).

Tang, Y., Xiong, J., Becerril-Arreola, R. and Iyer, L. (2019), 'Ethics of blockchain: A framework of technology, applications, impacts, and research directions'. *Information Technology & People*, 33(2), pp. 602–632.

Thompson, C. 2023. 'NFT Collective Proof signs with United Talent Agency'. Available at https://www.coindesk.com/web3/2023/01/07/nft-collective-proof-signs-with-united-talent-agency/ (accessed March 2023).

Ticketmaster Business. 2023. 'Ticketmaster launches token-gated sales, enabling artists to reward fans with prioritized ticket access and concert experiences through NFTs'. Available at https://business.ticketmaster.com/business-solutions/nft-token-gated-sales/ (accessed March 2023).

Turner, D. 2020. 'How private equity drained the record industry'. Available at https://www.getrevue.co/profile/pennyfractions/issues/penny-fractions-how-private-equity-drained-the-record-industry-244785?utm_campaign=Issue&utm_content=view_in_browser&utm_medium=email&utm_source=Penny+Fractions (accessed November 2022).

Turner, D. 2022. 'The record industry needs you to love NFTs (part 2)'. Available at https://pennyfractions.ghost.io/the-record-industry-needs-you-to-love-nfts-part-2/ (accessed November 2022).

UK Parliament. 2020. 'Formal meeting (oral evidence session): Economics of music streaming'. Available at https://committees.parliament.uk/event/15732/formal-meeting-oral-evidence-session/ (accessed November 2022).

Wiggins, K. and Nicolaou, A. 2022. 'Hipgnosis Songs Fund launches debt-funded buyback to halt share slide'. Available at https://www.ft.com/content/5625d351-868c-4644-b9b7-f36284553dfe (accessed November 2022).

4

MUSIC AND DIGITAL CREATIVITY

In May 2022, the European Commission announced that it had written to a number of European Union member states, requiring them to explain why they had not yet transposed into law the EU Directive on Copyright and Related Rights in the Digital Single Market (the 'DSM Directive'), a directive that had been ratified in April 2019 (European Commission, 2022). Article 17, a key component of the directive, was introduced in order to ensure that copyright holders receive due payment for the use of their content online (BBC News, 2018).[1] In the lead-up to its implementation into EU law, the DSM Directive had been the cause of considerable controversy. On one hand, some of its more notable opponents included Alphabet and Meta, owners of online platforms Google and YouTube, and Instagram and Facebook, respectively, along with the Polish government, who had challenged the directive through the European courts (Yun Chee, 2021). On the other hand, among its supporters were Universal Music Group and Warner Music Group, who, alongside bodies such as the UK-based Society of Authors and the Alliance for Intellectual Property and Proponents, were focused on protecting the interests of rightsholders and content producers (Reynolds, 2019). This split between the supporters of the directive – digital platforms – and its opponents – rightsholders – gives an overall shape to the perspectives shaping each side of the debate, but beyond these larger corporate differences it is possible to make out the contours of two fundamentally different ways of understanding creative practice. For the two music multinationals, Universal and Warner, for most of their existence, music has been a commodity; entertainment to be paid for, either as a one-off purchase, or as rental, via subscription to a streaming service. This was the traditional 'one-to-many' model that arose with radio and television, the mainstream broadcast media of the twentieth century. Although no less commercial, digital platforms such as YouTube, Facebook and Instagram, by design, operate

DOI: 10.4324/9781003225836-5

according to the 'many-to-many' model that emerged in the connected environment of the Internet, and has subsequently flourished in the age of the of the social web. As supporters of the DSM Directive, it would seem that rightsholders such as Universal and Warner have remained committed to defending a traditional model of value that exacts payment for songwriters, composers and other contributors in exchange for use of copyrighted digital content. Conversely the likes of Alphabet and Meta, its opponents, are aligned with an evolving articulation of value, where in addition to generating revenue from advertising and subscriptions, these platforms' focus is on creating social capital and indirect financial reward through the propagation and movement of online content.

We could think of these two models of content distribution as linear and exponential. The former conveys the sense in which content operates and should be paid for in one way, reflecting a view that music content is linear in nature. The latter is a recognition that, by its very nature, digital music content is hybridised and open-ended, as are the ways in which it can be used and paid for. In practice, both exist simultaneously. These are matters of ecology. Music, in its pervasive digital form, is both an ecosystem itself, where the production and consumption of music bring together a variety of activities into one entangled whole, and at the same time, it is part of multiple ecosystems, including technological, economic, social, cultural, scientific and environmental. Speaking in 2020, Spotify CEO Daniel Ek's response to critics of the streaming platform's royalty rates highlighted how music is on the faultline dividing these two visions of commerce, a divide spanning two centuries and two technological paradigms: 'you can't record music once every three to four years and think that's going to be enough [...] the artists today that are making it realise that it's about creating a continuous engagement with their fans' (Ek, in Dredge, 2020).[2] The new workflows for musicians that Ek describes, which draw on music's exponential growth within the platform environment, demonstrate that music itself is a multivalent phenomenon, which cannot be reduced to a single point of origin or function. As we have seen in previous chapters, music is fundamentally composed of multiple forces, and is irreducible to a single point of meaning or purpose. Despite strong opposition from both the music community (Cummings, 2020) and the music press itself (Beaumont, 2020), it would appear that Ek's views are aligned with the sense that music is fundamentally non-linear and entangled in nature, in terms of both its points of origin and its open-endedness.

Although contentious and polarising, Spotify's position on the changing role of the musician, and the DSM Directive's move to protect payments to rightsholders are clear indications of the extent to which digital platforms have disrupted established approaches to music commerce and traditional practices of music making. From TikTok's facilitation of mass amateurism and online production, to creative collaboration tools and live-streaming, in the early years of the twenty-first century, we have learned to make music in abundant and radically diverse ways. This chapter is a reflection on how participative and collaborative practices have evolved traditional working methods and frameworks in music. Equally, it is important to acknowledge how new distribution mechanisms are altering and challenging traditional conceptions of creative music in the context of professional practices. As revenue models change, and the everyday creativity and mass amateurisation models of the web 2.0 paradigm extend into their second decade, what has emerged is a clear view of several music industries operating in parallel. In the coming pages, I examine these changes, and explore how creative practices of producing, sharing and consuming music have evolved, by considering how software tools and network dynamics have facilitated new musical behaviours within the contemporary platform environment.

Interdependence and Determinism

There are established frameworks for thinking about cultural production in the context of technological environments, and the idea of technological determinism is one such framework. In Chapter 1, we saw how a connecting line can be drawn between the work of Karl Marx and contemporary materialist philosophies; both new materialism in its broad sense, along with what Galloway referred to as Laruelle's 'weird materialism'. Now, we can turn to Marx's work again, and understand how his proposal that human existence is structured by relations of production and conditioned by material – in other words, technological – forces of production (Marx, 1904: 11) underpins notions of technological determinism. On Marx's reading, the fact that these relations of production are independent of human will and constitutive of the foundation and economic structure of society (ibid.) drives the idea that, at a fundamental level, technology is the driving force behind all human history.

The relation between Laruelle's work and Engels' notion of 'determination in the last instance' is useful here, as it can help us develop a more nuanced perspective on what determinism might mean. Where theories of technological determinism place varying degrees of emphasis on the central role that technology plays in shaping – determining – the progress of human history, notably social development (Reed, 2018: 10), Laruelle's work enables us to displace traditional notions of determinism. Following Engels, Laruelle sees that the word 'determination' invokes a particular form of causality, one that emphasises the unilateral movement from the One to the world of lived experience. The concept of being 'given in' is thus a very particular reading of 'determination', one that disrupts traditional conceptions of causality, and which gives us the sense that – in non-philosophical terms – 'determination in the last instance' is synonymous with the idea that the world of real, lived experience is *given in* the absolute, non-accessible Real of the One. As a result, the non-philosophical approach breaks causal chains of determination, and the word 'determination' itself becomes a way to understand that, while the real world is given in the

One, and is determined by the One, the world is not directly *caused* by the One. We cannot say *how* things come into the world, because we can have no knowledge about the One. All we can say is that the world is in the One.

This would suggest that there is no absolute causal relation between technology and human history, simply that human history is given in the 'real' of technology (in the same way we saw that photo-fictions are given in the 'real' of photography). Although their models are radically different, the results of Laruelle's non-philosophical fracturing of causality can be likened to Barad's theories of mark-making and boundary-making. For Barad, causality – discerned via the presence of a mark on a body – is no more direct than Laruelle's notion of determination in the last instance. Marks on bodies, which bear the traces of boundary-making practices, do not result simply from an encounter with another object or force. Boundaries are formed via the measuring of one real body by another, and it is the boundary created by this measuring that creates the mark. Thus, Barad's work takes us closer to a sense of interdependence, wherein the entanglement of real human histories and real technologies create further, new reals; topological moments of measurement, the traces of which are left as marks on bodies. In this light, it is worth noting how the English and American Studies scholar T.V. Reed recognises a similar impulse in Marshall McLuhan's theory of technological interdependence, where, in a manner that anticipates Barad's new materialist inflections, McLuhan suggests that just as we humans shape our tools, so our tools shape us (ibid.: 10).

Whether we choose to think in terms of what Alexander Galloway sees as Laruelle's encrypted Real, which, as we saw in Chapter 2, arrives in our midst without any trace of its provenance, or an entangled real, more akin to the new materialist-inflected approaches of Barad, Morton and Haraway, my interest in this chapter is to further explore how relations between technology, practices of music, and music makers and listeners themselves are interdependent and co-determining. For non-philosophy and new materialism alike, the real is always in the world; we are always affected by it, and our actions are always fundamentally accented by a real that exists beyond our capacity to invent it. Where traditionally Marxist theories of technological determination emphasise how tectonic, technological forces operate at scales beyond our comprehension to ineluctably drive social development forwards, thinking interdependently is to recognise that things always bear the traces of their making, of a making process that has an objective truth that exceeds human subjectivity. As entangled practices, making, sharing and listening to music always speak of their own production, and simultaneously make marks and boundaries that are new sites of production and experience. Thus, while theories of interdependence challenge us to think more dynamically about the ways that practices of music making and listening might be entangled with a range of technological affordances and paradigms, the chapter also looks directly at how an evolving technological environment has offered opportunities for creative practitioners and audiences to engage with music.

Everyday Creativity

Published in 2011, David Gauntlett's book *Making Is Connecting* examined the flourishing of what he described as 'everyday creativity [...] a process which brings together at least one active human mind, and the material or digital world, in the activity of making something which is novel in that context, and is a process which evokes a feeling of joy' (Gauntlett, 2011: 221).[3] Gauntlett's focus was on how Internet cultures in the first decade of the twenty-first century – commonly referred to as 'web 2.0'[4] – had catalysed a new form of creative activity, and the book celebrated the mass creativity afforded by a new generation of online creative and communications tools. Essentially, Gauntlett's interest was in the way that media platforms such as YouTube, and creative tools such as GarageBand, were facilitating creative practices of making and sharing in the context of an accessible many-to-many communications network at a scale not known before. These ideas were reflected in a number of contemporary books, including Charles Leadbeater's *We-think* (2008), which similarly recognised the way in which mass-participatory projects such as Wikipedia and online games were demonstrations of people working together in new forms of association different from twentieth-century industrial hierarchies. Indeed, one of Leadbeater's key contentions was his sense that in the wake of the disruption caused by the rapidly evolving Internet of the early twenty-first century, which catalysed wholesale change in traditional industrial structures and behaviours, the everyday consumer was 'caught in the crossfire' (Leadbeater, 2008: 88). Clay Shirky used the term 'mass amateurisation' (Shirky, 2008) to reflect the scale at which web 2.0 technologies were enabling creative production, and to emphasise the non-commercial nature of this activity. Shirky drew on McLuhan's ideas relating to the evolution of media and production technologies – most notably the development of Gutenberg's printing press – and the potential of technological innovation to radically alter not only the relations of production, but also social relations. While all three writers made detailed articulations of the economic structures that underpinned this emergent creativity of web 2.0, there is nevertheless a certain utopianism in their views of the Internet in the late 2000s, and its capacity to foster cultures of sharing and making that are seemingly free of the strictures of corporate ownership, and of the need to harness creative activity in the service of earning a living. Utopianism aside, whether we choose to examine these developments within the scope of the new materialist tendency towards interdependence or the speculative realist/non-philosophical tendency towards determination in the last instance, these analyses not only provide valuable perspectives on an arc of change in Internet cultures that has developed during the last two decades, but they enable us to consider how the Internet's history of technological development has amplified, reflected and co-evolved with certain behaviours, expectations, social and commercial relations, as well as new means of production.

Although music has always existed online in a variety of ways, at the time of writing, there are three notable vectors that offer opportunities for musicians to create and share their music online that I explore in this chapter; the creator economy, the gamification of music and the social music economy. As emphasised in theories of everyday creativity and mass amateurisation, it would be a mistake to assume that commerce is either the ultimate or natural destination for musical creativity. However, as we saw in the last chapter, we would also be remiss if we fail to acknowledge, and explore, how the exploitation of intellectual property rights, in the form of music copyright, has changed, and continues to change exponentially in online ecosystems. Digital streaming platforms including Spotify, Apple Music, Tidal and Amazon Music have brought changes to music production and consumption patterns that include the proliferation of commercial and user-led play-listing, modifications to trends in composition, songwriting and structuring. Moreover, these innovations, alongside the revolutions in the management and exploitation of intellectual rights that we have explored in relation to song catalogues and blockchain technologies, have radically altered how musicians earn a living from music. Ultimately, the digital convergence of music and money as programmable information, has given rise to a new malleability for music; as an intangible data asset, music now offers an increasing variety of commercial opportunities to musicians, creators and rightsholders. In the last chapter, we explored how song catalogues and blockchain technology systems have proliferated in the digital music ecosystem, and engaged with changing practices of music making and listening that have arisen in relation to these technologies. In the coming pages, we engage with music's evolution as a social practice, exploring how new forms of music making and musical interdependence are being played out in the context of music as an online social phenomenon; a trajectory that has accelerated as the socio-economic environment of the Internet as a mass communication ecosystem has itself evolved beyond the scope of what Shirky, Leadbeater and Gauntlett were describing in the first decade of the twenty-first century.[5]

The Creator Economy, and a Creator Ecology

In the context of a contemporary Internet culture which has fully acknowledged and embraced the commercial potential of 'everyday creativity', in this part of the chapter I want to address everyday practices of making and sharing in terms of what has become known as the 'creator economy'. According to the business magazine *Forbes*, 2021 marked 'the beginning of the end' of the era of the Attention Economy – an economy based on advertising revenue – which has given way to the Creator Economy, an economy it describes as being composed of 'platforms, marketplaces and tools [that are] democratising creative expression and entrepreneurship' (Bergendorff, 2021). The fundamental difference that Bergendorff's article describes is the paradigmatic shift away from an economic model underpinned by the selling of vast quantities of consumer data, towards a situation where the revenue paid directly to creators across a variety of platforms is shared with the platforms themselves. Describing creators as 'micro-entrepreneurs' who often operate in niche markets and who build up intimate knowledge and understanding of their patrons' tastes and appetites vis-à-vis content and products, Forbes' analysis showed that, in 2021, the creator economy was a hybrid of the creative industries and the gig economy,[6] with 14 per cent of gig economy workers operating in the creative industries (the highest percentage across all sectors).[7]

According to MIDiA's analysis, in 2022, the number of creators independently releasing their music on streaming platforms increased by 12 per cent, whereas the number of creators uploading and sharing their work on other platforms increased by 21 per cent. As of 2021, there were 64.1 million music creators who were producing and/or recording their music, a figure which is set to rise to 223.4 million by 2030. While there are no exact figures for how many of these millions of creators are earning a living from their creative activities, average earnings for a full-time professional music creator stand at $31,268, and for a part-time professional, $5,128 (Cirisano, Mulligan and Thakrar, 2022). MIDiA also emphasise how a prime focus of the most recent wave of innovation in the creative audio software sector has been on enabling its target audience to earn money. Such a focus is also an acknowledgement of how the entangled nature of the creator economy not only involves digital platforms, musicians and music consumers, but it also integrates music technology companies. Understanding how the music technology sector is playing a key part in the evolution of creative music practice is to recognise an interdependent relationship that involves platforms, creators and technology companies. From the creators' perspective, they are able to generate further revenue by using a platform's products in their own creative endeavours, audio productions, skills sharing videos and other new marketplaces. This in turn also serves to widen a platform's renown as a creator's works spread throughout connected communities. For the audio technology sector, by incentivising users to earn money by using their products, and reinforcing the utility and value of their branded audio equipment, companies increase the likelihood that their customers will spend more on their products and services in the future.

In the UK government-commissioned report *Music Creators' Earnings in the Digital Era* (2021), David Hesmondhalgh et al. define music creators as 'those involved in performing for recordings, composing music and songs intended to be recorded [and] studio producers, sometimes called record producers' (Hesmondhalgh et al., 2021). The defining characteristic each of these roles shares is that they are 'integrally involved in the creation of recordings or songs that are [...] eligible for protection under copyright law' (ibid.). This is to say that, for the purposes of the report, music workers who create copyrighted material fulfil the criterion of 'music creators': 'in our view, all music creators are musicians. But not all musicians are music creators in the sense we are using the term here: some play music but do not fix their music in copyrightable form' (ibid.). Along with Forbes, MIDiA categorise music creators very differently from Hesmondhalgh et al., identifying three main sources of income: producing/mixing /engineering, streaming and performing (Cirisano, Mulligan and Thakrar, 2022).[8] The Hesmondhalgh perspective is useful as it provides us with a particular emphasis on how musicians are seeking to generate income from their musical labours, but it is important to note that, in terms of how music creators in every sense earn an income from their activities, it is increasingly understood that the streaming-centric model cannot sustain success in the long term. This is an important point about the divergent identity of the creator in 2022, which again reflects Daniel Ek's point that the traditional model of the musician who makes a career from pure music making is now all-but obsolete. Thus, many income solutions in the creator economy, such as 'sounds and skills marketplaces, live streaming, NFTs, and production music', no longer operate according to an intellectual property-based revenue model (Cirisano et al., 2021). So while the Hesmondhalgh report's delineation of music creators does not match with Forbes' and MIDiA's definitions, nevertheless, it does enable us to draw a distinction between those who create revenue from licensing their copyrighted material, and those who, in the context of the recently emergent creator economy, create revenue in other ways. As such, where Chapter 3 focused on an ecology of rights-based revenues, this chapter examines the interdependence of creative music practices in the context of the Forbes/MIDiA creator economy model.

Through the early 2020s, as the creator economy became further entrenched as a means to simultaneously generate greater revenue through greater creator–consumer engagement, digital platforms began to launch what they called 'creator funds'. An article posted on the Shopify Blog in November 2021 listed 13 such funds, including TikTok's Creator Fund, the YouTube Shorts Fund, Snapchat's Spotlight Challenge, Facebook and Instagram's Creator Program, along with a raft of funds run by the likes of LinkedIn, Spotify and Pinterest (Adegbuyi, 2021). Underpinning these funds is the idea that, by investing in high-performing social media influencers, platforms can achieve mutually beneficial growth: growth of an influencer's audience and of the platform's engagement figures. Describing this as a 'symbiotic' relationship, Adegbuyi discusses how the concept of investing in influencers so as to incentivise them to generate more engagement on native platforms was a deliberate move away from what would have been the norm at the time of *Making Is Connecting* where content was largely made for free, and if sponsorship was involved, it would be in the form of a partnership for third-party brand endorsement (ibid.). The creator economy is not without its critics, however. Established in 2020, with an initial outlay of $200 mn, TikTok's Creator Fund has been seen to deliver diminishing returns to its creators. Unlike YouTube, who, through its Partner Program established in 2007, uses a revenue-share model to apportion revenue from advertisers whose adverts are shown at the beginning, during or at the end of a video, the TikTok fund simply pays out from the initial static sum of money (Silberling, 2022). The upshot of this situation is that, whereas YouTube creators are paid in line with YouTube's own growth – in other words, the more advertising revenue a YouTube video generates, the more a creator will get paid – as more TikTok creators enter the marketplace, each earns less as the fixed size of the fund is shared with a growing number of creative contributors.

The Forbes article divided the creator economy into four categories that were aligned with potential investor opportunities: platforms, tools, monetisation enablers and creator-led businesses, suggesting that many companies combined elements from each area (Bergendorff, 2021). In the context of theories of interdependence and determination, acknowledging how a particular platform environment can inform and emphasise certain approaches to working - and how a given set of creative tools can influence the type of products that are being produced - is essential to understanding and engaging with the features and behaviours that characterise current digital ecosystems. For musicians, the platform environment is both a social and professional space, and Bergendorff lists a number of features that characterise platforms that are relevant to music creators, including a platform's capacity to democratise access to audiences, and to enable creators to generate a variety of digital entertainment beyond traditional formats. What is more, because of network effects, users can become increasingly locked into a platform as their content generates revenue (ibid.). This is to say that content may increasingly accrue value because of its relation to a specific platform, rather than holding inherent value in itself: musical content on social networking platforms such as Instagram and TikTok being obvious examples of music having an on-platform value that may not reach beyond the confines of a particular platform's environment. We can also understand how platforms interact with musicians at the level of behaviour and identity, and Bergendorff suggests that, due to the proliferation of easy-to-use digital tools which subtly transform amateur creative activity into commercial production, the platform environment catalyses the 'enterprisation of individuals' (ibid.). Thus, musicians working in the creator economy become 'micro-entrepreneurs' and develop 'micro-brands' across a raft of platforms offering an increasingly accessible production and distribution pipeline. For Bergendorff, these changes are further amplified by the third and fourth categories of the creator economy. The recent rise of NFTs exemplifies the evolution of monetisation enablers supported by the widespread proliferation of online marketplaces, and novel payment solutions, for example, in the form of in-app purchasing, which we shall further examine below. In terms of creator-led businesses, these represent a new form of business, one that is native to the platform environment. He notes that such business operate by fully exploiting the platform economy infrastructure, a process that involves maintaining high consumer engagement by deploying a range of monetisation strategies (ibid.).

In his formulation of determinism, Althusser described these types of internalisation of ideological values and behaviours as 'interpellation' (Althusser, 1970), a phenomenon that we might recognise in the sense that music creators are subsumed by the platform environment's capacity to alter their behaviours and individual value systems. To think interdependently is to recognise such changes, but it also involves recognising the hybrid and unfixed nature of platforms themselves. Just as we see musicians and music practices changing and evolving in relation to the platform environment, so too do platforms evolve, as we can see from the shift from web 2.0's culture of amateur making and sharing, where platforms created value through advertising and increased user engagement, through to the creator economy of the early 2020s where platform–creator relations are far more complex and intertwined.

'Prosumerism' has long been associated with the everyday creativity ethos of the early twenty-first century. Derived from the portmanteau 'prosumer', the concept was coined by Alvin Toffler to articulate how production and consumption are not fixed categories, and that, depending on context, individuals can be both producer and consumer (Toffler, 1980). More recently, MIDiA Research have signalled a further shift: towards an economic model that aligns consumption with promotion. Writing in 2020, Mulligan and Jopling describe a continuum of digital platforms distributing huge swathes of content, providing consumers with the opportunity to engage with content in various ways. Another way to think about this continuum is as an ecosystem of platforms, interdependently influencing and responding to each other; a creator *ecology*. In this ecology, creators can monetise their content in a number of ways, through advertising revenue, subscriptions, ticket sales or merchandise. At the same time, micropayments, virtual goods and crowd funding have emerged as alternative payment methods (Mulligan and Jopling, 2020).[9] What is more, because of the dynamic interrelations between what Mulligan and Jopling refer to as 'technology and content communities' – in other words, platforms and rights holders – the mass proliferation of cover versions, karaoke and lyrics have also generated further revenue opportunities (ibid.). Although our focus in this chapter is on how practices of music have evolved in the context of the creative web beyond a traditional transactional model based on royalty payments, these developments are noteworthy, as they indicate that even traditional payment systems are themselves evolving: on one hand, user-generated content presents a revenue stream for music publishers and collective management organisations that is potentially larger than the publishing revenues traditionally collected for recorded music; on the other, the technological providers and innovators within this creator ecology are coming to understand and adapt to the way in which copyright operates, particularly in relation to the complexities of rights and attribution, and global variations of licensing agreements. Mulligan and Jopling clearly understand that the rightsholder industries and technology industries each have different cultures, values and objectives. They see that music rights holders, which include labels, publishers and collective management organisations, have become what they call 'digital first' businesses. This is to say that many companies have created technological infrastructures to support a digital operations approach, including licensing teams and data analytics capabilities (ibid.).

An example of the emergence, rapid and significant growth, and concurrent evolution of the creator economy is the membership platform Patreon, which enables fans to support creators through regular monthly payments. In return for their patrons' regular payments, creators produce a range of original content, including music, videos, photographs, images, artwork, graphics, audio clips, comments, data, text, software and scripts. Jack Conte co-founded the company in 2013 with business partner Sam Yam as a means to generate more than the $50 per month he was receiving from his 150,000 subscribers and 1 million views on his YouTube channel (Hern, 2018). Conte's original vision was to persuade a few hundred of his most ardent fans to commit to becoming a regular patron, instead of trying to recoup money from the thousands subscribing to his channel or the millions watching his videos. By 2014, his videos were generating $7,000 from his patrons. By 2018, Patreon had 100,000 registered creators, who were being supported by 2 million patrons. It had paid out over $350 mn since 2013, and payments topped over $300 mn in 2018 alone. In 2021, Conte confirmed that Patreon had grown again. Over 200,000 creators were making money from approximately 7 million patrons, and Patreon was making payments totalling over $1 bn per year (Conte, in Hyman, 2021). In the last chapter, we engaged with music as a hybrid, intangible asset. What the success of Patreon demonstrates, as evidenced by the huge diversity of the content and paraphernalia that its creators produce, is the hybridity of the creator economy itself. Again, as a creator ecology, its hybridity speaks of the unfixed nature of creator platforms, consumers' appetites and interests, the increasingly open-ended role of musicians-as-content-creators, and the various forms that music and music making takes. Perhaps the most signal consequence of the creator economy is therefore that it reveals how seemingly fixed categories such as 'musician' and 'music' are but moments of intersection and entanglement. Turning to theoretical frameworks, we can see that from an agential realist perspective it is in the processes of producing these categories, rather than in the categories themselves, that we can identify an objectively existing reality. A non-philosophical and speculative realist interpretation would suggest that the real, absolute qualities of categories such as music and music maker are not the things that we can identify and understand; what makes them absolutely and objectively real is precisely that which we can never apprehend or understand.

In addition to Patreon, other companies exemplify the entangled nature of the creator economy. MIDiA reflect on how a combination of demographics and alternative revenue models for music-related creative labour is compensating for, and altering, the economic landscape for those wanting to create revenue from their music activities. Since 2010, a generation of audio platforms have emerged, such as the cloud-based collaborative music-making platform Splice, along with BandLab and LANDR, which are disrupting the music production ecosystem in two key ways. First, such platforms are being developed to align with the needs and requirements of a younger generation of musicians who are interacting with creative tools in ways that are very different from those of their forebears, not least because of their now relatively long-standing integration of AI-based tools. Second, in time, these changes to workflows look set to suffuse the entire audio production ecosystem (ibid.). Essentially, the message that comes across here is that demographics are central to how the music production economy and ecosystem are being reformed. The inference is that, given that the emerging generation of music creators are not wedded to an older model of rights-based income, then production software is evolving to maintain alignment with their changing needs and desires. Along with Hesmondhalgh et al. (2021), MIDiA notes that, while the streaming economy has opened up international markets to a range of professional and semi-professional creators to an extent that was unknown in the pre-digital era, the revenue model for streaming does not facilitate a sustainable living from recorded music alone. Although events like the DCMS Committee's Inquiry into the Economics of Music Streaming in 2020 have sought to address the imbalances in the streaming system, MIDiA suggests that 'no amount of increased royalties [will] fix the underlying dynamics' (ibid.). Ultimately, it would appear that tangible change will come via the development and adoption of a parallel and qualitatively different economic model for music commerce. For MIDiA, this change would take the form of new types of platform, for example BeatStars, a 'creator tools marketplace', whose subscription-based model enables producers to sell their music via a variety of exclusive and non-exclusive licensing arrangements (ibid.).[10]

In August 2022, Music Business Worldwide (MBW) reported that ByteDance, owner of the short-form video platform TikTok, had made a US trademark application for 'TikTok Music' (Stassen, 2022b). As with Epic Games' purchase of Bandcamp, ByteDance's development of TikTok Music reflects the growing awareness of opportunities for expansion and diversification offered by the integration of music with markets and technologies that go beyond traditional strategies for music distribution and commerce. The TikTok Music trademark application built on ByteDance's March 2022 launch of 'SoundOn', its proprietary music distribution and promotion platform operating in the UK, the United States, Brazil and Indonesia. SoundOn enables artists to upload their music to TikTok and Resso, the 'social music streaming app' that ByteDance launched in 2020 (Singh and Lunden, 2020), and which currently operates in India, Brazil and Indonesia (Stassen, 2022c). In addition, via SoundOn, artists are able to distribute their music to Spotify, Apple Music and Instagram (ibid.). Notably, TikTok's interest has been in facilitating the profiling and distribution of independent artists (in other words, artists who have not signed a contract with a traditional music publishing and distribution company), which again reflects MIDiA's articulation of the increasingly hybridised nature of consumption and promotion. The same MBW article also reflects that, with the launch of SoundOn, ByteDance have also opened up revenue opportunities, not only for themselves, but for the music industry more widely, particularly record labels. Indeed, the article reports that in 2021, from an overall revenue of $30 bn, YouTube contributed $6 bn to the music industry (Dalugdug, 2022). While

SoundOn is in its nascent phase, and has yet to create anything like these levels of income, given its global user base,[11] TikTok is seen as having the potential to create significant revenue for the music industry both through advertising as well as through the TikTok Music platform.

In November 2022, MBW reported that the three major global music companies – Warner Music Group (WMG), Sony Music Entertainment (SME) and Universal Music Group (UMG) – whose licensing deals with TikTok were come up for renewal in early 2023, were in negotiations with the social media platform to change the terms of their contracts. Since 2021, TikTok's partnerships with the music companies had been on the basis of buy-out arrangements, whereby the platform had made an upfront payment to its label partners who licensed the use of music for an agreed time period, in this case, two years. According to MBW's analysis, TikTok's growth is directly related to the use of music on the platform, which is thought to be limited to videos that are shorter than 30 seconds, and which serve to promote music on subscription platforms such as Spotify and Apple Music (Stassen, 2022d). Changing to a revenue share deal would align with Meta's arrangements with UMG and WMG. In a revenue-share model, the platform pays the rightsholder a proportion of revenue in accordance with an artist's or label's popularity on that platform.[12] In October 2022, Meta announced that its 'Music Revenue Sharing' strategy was being implemented worldwide, meaning that video content creators – for example creators of recipe videos – would be able to choose from an extensive catalogue of licensed songs to soundtrack their productions (Meta, 2022).[13] As we saw in the case of the TikTok Creator Fund, fixed funds can start to cause tension if the number of payable licensees starts to expand while the available pool from which the licensees can draw payment remains static. It is not surprising then that, according to MBW. ByteDance is being encouraged by global music industry leaders to adopt a similar revenue-share model for TikTok, which would establish a more long-term operating strategy for music licensing on the platform (Dalugdug, 2022).

Thinking in terms of a creator ecology equips us with a way to understand the contemporary music environment, and a more dynamic way to think about who is making music, who they are making that music for, and how that music is being consumed. In this context, it becomes increasingly clear that the idea of the 'musician' as a fixed entity with permanent attributes around behaviours, skillsets, their relationship to an audience, the nature of that audience, and deeper issues of political economy - in terms of the financing of musical activity, and therefore the purpose or at least hegemonic context for musical creativity - is also evolving rapidly in relation to a wide-ranging set of technological and economic changes. To give further definition to music's interdependent relationship with a set of production and distribution environments, I also want to map two other contexts for music practices, both of which can be linked to the creator economy, but are notable in their roots are in the games economy.

Music and Games: A Culture Change

In March 2022, Epic Games acquired the music platform Bandcamp (Stassen, 2022a). As the owner of Fortnite, Epic Games already had considerable experience of integrating the worlds of online gaming and music, and this acquisition is perhaps the most visible example of the increasingly interdependent relationship between the games and music industries, although the nature of this relationship has itself undergone rapid transformation. The company launched the concept of in-game concerts with the landmark live performance by Marshmello in 2019 that registered 10.7 million people in attendance. At the time, this was Fortnite's largest in-game event to date (Webster, 2019), and its success has led to an ongoing series of live concerts featuring the likes of Ariana Grande and Travis Scott (Regan, 2021). Fundamentally, what is most significant about Epic Games' acquisition of Bandcamp is that it represents a visible recognition, on the part of a technology-first company, of the opportunities of engaging with a content-first approach to the music sector. As MIDiA had noted in 2020, the challenge for platforms operating in the emerging creator economy has been to develop new features within the digital music and media landscape and thereby encourage new patterns and behaviours in terms of consumption (Mulligan and Jopling, 2020). By acquiring Bandcamp, Epic Games have enabled themselves to circumvent a 30 per cent fee that Apple Inc. apply to any purchases made within its proprietary App Store. So, while Spotify users who have subscribed to Spotify via the App Store have found themselves paying 30 per cent over Spotify's monthly subscription fee, since Apple apply that fee to Spotify for using the App Store (Carr, 2021), Bandcamp charges the artists who sell their content and merchandises via its platform between 10 and 15 per cent, which it waives completely on 'Bandcamp Fridays' (Music Business Worldwide, 2022). Understanding the repositioning of technology and content companies within the creator ecology is essential to recognising how the role and identity of the musical artist have also rapidly evolved within the online music ecosystem. It is telling that, as of 2020, artists already stood to make less by licensing a song for use in Fortnite than signing a deal with Epic Games themselves that would enable game players to buy 'audio moves' and other 'premium content experiences' (Mulligan and Jopling, 2020). What is apparent from the market analysis is that the in-app purchase paradigm that is native to the game ecosystem – the likes of Fortnite, GTA, Minecraft, Call of Duty – looks set to be an ever more important and valuable field of operations for music creators and the 'content first' community.

Music and Games: Live Streaming

While the technology to support live streaming has existed for 20 years, as a means of engaging with audiences at scale, it has been live streaming's gradual adoption by various social media platforms – the likes of YouTube, Instagram Live, Twitch and Facebook Live – where significant progress has been made during the past decade. Although Mark Daman Thomas notes that the singer-songwriter Sandi Thom had experimented with live streaming in 2006 (Daman Thomas, 2020), in general the early history of live streaming was largely as a broadcast medium for gaming. Launched as TwitchTV in 2011 (Business Wire, 2011), the live streaming platform now simply known as Twitch was originally focused on developing a broadcast network for games and esports. In its early stages, Twitch registered 3.2 million visitors per month. By 2012, site visitor numbers had increased to 20 million unique visitors per month (Geeter, 2019), and by 2014 to 55 million per month, prompting Twitch's acquisition by Amazon (Gittleson, 2014). Musicians have increasingly turned to Twitch as a means to create new opportunities to connect with fans, and during the global COVID-19 pandemic, the platform's viewing figures for live streamed music events grew by 550 per cent (Sisario, 2021). While Twitch is far from alone in the live streaming space, for anyone spending time exploring the available streams in its Music directory (Twitch, 2022), it quickly becomes apparent that the architecture of the environment, with its live chat stream, and the list of recommended channels, draws on gaming, rather than music, archetypes. This would suggest that, again, what might have previously counted as a native environment for music-based activity has been rapidly transformed by the interests of a changing audience, alongside the opportunities afforded by technological innovation, and monetisation processes.

Although the public lockdowns that came into force during the pandemic clearly caused a huge spike in online activity of all kinds, in a post-pandemic climate, audiences for live streamed music will remain significant. MIDiA Research notes that, alongside the likes of Twitch, YouTube Live and Instagram Live, smaller platforms have continued to flourish. With record attendance figures for live streamed concerts – for example, 750,000 paying attendees for BTS's Bang Bang Con, and figures in excess of 100,000 for Erykah Badu's live streamed shows – it would seem that live streaming constitutes a new sector of the music industry, with the capacity to reach new audiences (Mulligan and Jopling, 2020). Live Nation, the global entertainment company whose subsidiaries include Ticketmaster and Festival Republic, operate a live stream portal that enables online audiences to stream a number of its in-person concerts (Live Nation, 2022). MIDiA's interest lies in the capacity of the live stream format to offer musicians the flexibility to 'innovate with more immersive and interactive formats', which could include experimenting with different video and sound feeds, and to try out material that might not work in standard live environments (Mulligan and Jopling, 2020). In the context of the role that live streaming is playing in the increasingly hybridised creator economy, it is also important to reflect on how the format offers a range of revenue opportunities for creators, including income from performing rights, advertising, tickets and subscriptions (ibid.). MIDiA reported in 2021 that live streaming had become a recognisably lower cost means of accessing large concerts and sporting events, noting that the sports sector had already begun expanding its digital offer to enable fans to engage with events at a distance. Not only are there a number of economic imperatives here, in that both the sports and music communities are working to develop ways to increase their audience share via a digital offer, but in terms of the wider ecology of live events, such changes have begun to reshape our expectations of what a live event is. From an environmental perspective, if we can have a high-quality experience without having to travel to see a band, then there are benefits in terms of a concert's carbon footprint. However, in more subtle ways, if bands and artists begin to expect that a sizeable proportion of their audience will be joining them online, rather than in person, that also starts to change the nature of the event itself. In order to satisfy the needs of a hybrid audience, then, a hybrid live event will doubtless need to evolve from what used to be a completely in-person event.

While this is still very much a developing narrative, what these changes to the infrastructure of music distribution and consumption suggest is that the most notable impact on music from games cultures may be less focused on the gamification of music itself, and more focused on a set of behaviours that musicians and music audiences increasingly adopt. As music becomes further embedded within the games economy, where economic exchanges and commercialisation opportunities operate very differently to those in the music streaming environment, the recalibration of the music economy in terms of the opportunities offered by the games ecosystem may bring music to a point of radical socialisation.

Social Music

Founded in 2015, BandLab describes itself as a 'social music platform that enables creators to make music and share the creative process with musicians and fans' (BandLab Technologies, 2022). So far in this chapter, I have looked at how certain key trajectories in music – revenue generation, audience development, a changing market for music as a cultural product – have adapted and changed as technology and cultures have undergone radical transformation since the heyday of web 2.0 in the early twenty-first century. With the arrival of platforms such as BandLab, it is possible to discern a technology-first music company operating in a manner that is native to the socially focused world of the contemporary Internet. The BandLab homepage goes on to describe how 'BandLab combines music making and collaboration tools like the world's first cross-platform DAW, with social features like video sharing, messaging and discovery' (ibid.). As we saw with Twitch, which has a set of user features that go far beyond the usual format of streaming platforms and push the experience and means of consumption for live streamed music far closer to what has been traditionally the domain of online gaming and Internet chatrooms, BandLab has set out to create a music-first infrastructure that is native to a highly interconnected and interdependent world; interconnected, because its features are a recognition of our networked digital communications environment; interdependent, because the platform's design is an acknowledgement that for growing numbers of amateur and professional creators – particularly in the light of MIDiA's predictions for the number of music creators likely to be active by 2030 – having the ability to share and to communicate content is now as important as having the ability to create content. Launched in 2022, Elk LIVE is a system that combines an audio interface with an Internet bridge – the proprietary Elk Bridge – that enables musicians who are up to 620 miles apart to play together in real time with no latency; in other words, there is no delay between the musicians hearing each other (Elk, 2022 and Hall, 2022). Such systems are not entirely new, but latency has been a perennial challenge for any developers wanting to work in this area. NINJAM, which launched in 2005, enables geographically distributed musicians to play together by putting a musical context around its system's latency. The software allows users to specify a set number of bars delay between when they hear their fellow musicians playing, and when they themselves play (Cockos Incorporated, 2022). For example, if two musicians set a 4-bar delay between them, then the second musician will hear beat one of the first musician's first bar when the first musician is actually already at the start of their fifth bar. It's a system that can be adjusted to with practice, and the experience can bear some resemblance to the way that canons or musical rounds are constructed.

Systems like BandLab and Elk certainly suggest that inherent in the desire to make music is the desire to make connections; and it would seem that, as an ecology, music is currently the site of a new entanglement between creativity, sharing, technological innovation, socialisation and monetisation, wherein traditional modes of music practice are being recalibrated. Through a series of creator economy analyses in 2021 and 2022, it is clear that MIDiA see a number of fundamental changes occurring in the creative music-making sector. There are increasingly more creators – in other words, non-specialist music makers – who are making music, many of whom do not play music in a traditional sense. A number of platforms are now offering in-house music creation opportunities, which offsets the need to learn to play an instrument or create music in a traditional way (including TikTok, Roblox, Fortnite, LANDR and Splice). As software develops, it is becoming easier for creators to produce original content, rather than having to repurpose existing material (this is particularly true of AI-enabled platforms such as LANDR, which I discuss in Chapter 5). The growing creator class appears to be less concerned with high-level production values than with 'creating with and within communities of like-minded creators' (Mulligan and Thakrar, 2021 and Cirisano, Mulligan and Thakrar, 2022). This re-alignment and re-balancing of the music ecology, indeed, with the relatively recent arrival of new component parts of the ecology in the form of creation-for-communities, undoubtedly means that the practice of making music, and many of the traditional expectations that surround it, are also changing. Debates surrounding Spotify's 30-second threshold for activating streaming royalties have a long history (Vincent, 2015 and Richards, 2022), including the impact of this mechanism on songwriting in terms of structure and aesthetics (Mack, 2019). The way in which music is becoming entangled with an entirely new set of forces and trajectories is significantly increasing both the nature and scale of the change to familiar forms of creative music practice.

Audiences and Consumers

Looking at music from the perspective of its makers and the various production and distribution channels has painted a vivid picture of just how unfixed and open-ended the practice of making music remains. However, thinking about audiences for music in the creator economy will provide us with an even more comprehensive picture of music's ecological and dynamic make-up. As of 2022, it would appear that even amongst a music-first audience, the act of listening to music as an activity in and of itself is being given less and less time. As more entertainment hours are allocated to watching video and engaging with online social activity, the artists themselves are focusing more of their attention on platforms that will enable them to create revenue from their content – by monetising their fanbase in various ways – and dedicating less time to streaming (Mulligan, Thakrar and Cirisano, 2022). In the same way that entire generations have never engaged in audio cassettes or compact discs, a new generation of music consumers are less interested in the music-first offer of streaming platforms, and are instead more focused on the active environments offered by social-first platforms. Two points arise that strongly indicate the ecological nature of music within the online social environment. Firstly, MIDiA show that only 33 per cent of consumers now only 'passively' consume online content, which is to say that they simply listen or watch it, without taking any other actions. This compares with 18 per cent of audiences that are also content creators themselves (ibid.). Music creators and music consumers are therefore part of the same continuum in a way that is very much native to the current digital ecosystem. This is to say that producers and consumers are both engaged in the same kinds of creative activity, albeit with different objectives in mind, and no doubt operating with different levels of skill and specialist knowledge. However, this means that, increasingly, the divide between creator and consumer is becoming blurred, which in itself changes how a music maker approaches the task of making music. Questions that might have once been relatively straightforward to answer have become complex: Who am I making music for? Who is my audience? If creative audiences are increasingly wanting to experience music that can be repurposed in various ways, and consumed in a variety of non-audio only contexts, then new challenges have arisen that music makers are having to acknowledge. The second point is that, with the arrival of 'the platform era', a transformation in artists' behaviour is underway, triggered by the globally – and therefore constantly – accessible nature of platforms (ibid.). As the backlash against Daniel Ek's pronouncements about the new workflow for artists in the platform era has demonstrated, some are unwilling to adapt to the demands of the platform economy, seeing that time spent courting their social media fanbase is time that could be better spent creating new material (Beaumont, 2020). These are indeed specific changes brought about by the onset of the platform era, which has facilitated a new set of relationships between listeners and consumers and the music and musicians they follow and want to support. While these changes are evolving rapidly, we must remember that music has always been ecological; the platformisation of music is but the latest iteration of change. And taking a broader view, it is important to recognise that music has never been in one fixed mode of consumption, neither has it been produced with one fixed purpose in terms of how people will come across it. Music as a communal experience is now finding new modes of social interaction, and in this sense, this new sociality is simply an overt demonstration of music's ecological nature. If digital sociality is the new environment, or at least the new direction for music making and consumption, then this would suggest that music is nothing, if it is not something that is part of many other things.

One of the most compelling statistics that MIDiA have produced is the growth in what we might describe as 'social-first' music makers, relative to 'streaming-first' musicians. Although Hesmondhalgh et al. (2021) restrict their definition of music creators to those who make a living via revenue raised against their original recordings on streaming platforms, and the 'creator economy' is an all-encompassing term that takes in a number of revenue opportunities offered by the contemporary platform economy, perhaps 'social-first' and 'streaming-first' can work to capture a fundamental faultline and point of change between two states of what it might mean to be a musician. MIDiA show that the number of musicians releasing music on streaming platforms via a distributor rose 12 per cent from 2021 to 2022 to reach 5.6 million. The number of musicians posting directly online (which includes TikTok, Shorts, YouTube, SoundCloud, BandLab, LANDR) grew by 30 per cent in the same period to reach 31 million (Cirisano, Mulligan and Thakrar, 2022). These quite stark figures, and the clear increase in social-first activity relative to streaming-first activity, demonstrate the extent to which the various forces that we have looked at – the exponential socialisation of music making and music-related activities, the hybridisation of the games and music economies, and the diversification of the creator economy – are radically transforming perceptions of what making and experiencing music might be. While the word 'perceptions' might suggest that we are conveniently dismissing the various realist positions that underpin the trajectories of this book, Karen Barad's work reminds us that these changes are real. Perceptions as instances of agential realism – in other words, a point of measurement of a real occurrence in the world that itself creates a real outcome – mean that the digital socialisation of music, with all its concurrent shifts in expectation around form, purpose and content, constitutes a tangible change that is happening in the ecology of creative music practice.

Conclusion: The Creator Ecology… and a Music Industry that Does Not Exist

We are at a curious moment in the history of music. According to the website Earthweb, in 2022, over 600 channels on YouTube have more than 10 million subscribers (Wise, 2022). However, despite the renewed popularity of her 1985 hit, 'Running Up That Hill', thanks to its appearance on the television series *Stranger Things* (Netflix, 2022), at the time, the musician Kate Bush had only 688,000 subscribers (KateBushMusic, 2022). The current configuration of the creator economy begs the question about whether a YouTuber with 10 million subscribers is any less professional than a musical artist with 688,000 subscribers who is signed to a record label, and whose music is featured on a major Netflix series. While this is not an entirely serious question, it does illustrate the nature of the ecology in which successful YouTubers and successful artists like Kate Bush operate, and encourages us to think about what the actual differences and similarities are between these kinds of creators. David Gauntlett emphasised how everyday creativity was a phenomenon that was powered by the accessibility of creative tools and the means to express that creativity at scale, where, for example, an amateur YouTuber making video content in their bedroom and Beyoncé have the same potential audience on YouTube, even if their videos aren't generating the same viewing figures. In essence, a key outcome of Gauntlett's research was his framing of social making in the online world as something that resulted from the common capacity to do something that a person or group had not done before, harnessed to the capability to share the fruits of that creative endeavour more widely than at any point in human history. While the term 'creator economy' is suggestive of the ongoing financialisation of the web, and of the democratisation of the opportunity to generate an income from online activity, clearly, its roots lie in the far more amateur-led environment that Gauntlett and his early twenty-first century colleagues were exploring. Indeed, even as recently as 2020 MIDiA were using the term 'mass customisation' to describe how consumers, in producing user-generated content, were deepening their engagement with online content by 'personalising it [...] sharing it with a cover note, liking it, adding music to photos, creating a dance video to a song or even creating their own versions or mash ups' (Mulligan and Jopling, 2020). However, I would suggest that the fundamental insight that we can take from the evolution of content production and consumption that we have tracked, is that creators, content, platforms and consumers are all aspects of an interdependent creative ecosystem. Music, musicians, and listeners are all part of an evolving ecology, where the differences between user-generated content and professionally generated content are increasingly differences of degree, rather than differences in kind, where the differences between a professional musician and an amateur creator are no longer defined by the once legitimising function of a management contract, a publishing deal, or a recording deal. Indeed, it appears that professionals and amateurs are now more similar than they are different, by dint of their immersion in a platform ecology that consists of multiple pathways to building and connecting with an audience, to generating revenue, and to creating new work or variations of existing work.

The likes of TikTok, YouTube, Instagram and Snapchat have developed revenue models and opportunities that are fundamentally different to the rights-based model that drives the streaming economy. The creator economy operates in real time, tethering creator income to a platform's capacity to attract advertising revenue, and therefore the nature and purpose of the work undertaken within the creator space is very different from work undertaken to meet the demands of the streaming space. In the context of the creator economy, music-based activity – live performance, reviews, reactions, comedy skits, short-form videos – has become the socially rooted site of music consumption and reframing of musical production. It could be argued that the performative aspects of the creator economy are akin to street performance or busking, where an audience is gained and won over, and an income is earned by virtue of a performer's capacity to keep passersby engaged. This may be the case, and, if so, then the multi-skilled approach of the street performer – who must carefully breathe fire or swallow swords whilst at the same time reaching out to attract the attention of the next passerby to walk on to the town square – is an approach that is being increasingly adopted by digitally social musicians.

Hesmondhalgh et al. (2021) make a number of salient points about the current state of the digital music economy. Their report describes the increasingly competitive and over-saturated nature of the streaming market, and notes that in 2019, 'combined UK revenues from on-demand streaming, downloading, physical sales, synchronisation licensing, and public performance and broadcast rights were worth approximately £1.69 bn', compared with a figure – equivalent to 2019 prices – of £1.99 bn in 2000 (Hesmondhalgh et al., 2021). However, acknowledging this sizeable fall, the report anticipates that, in the coming years, revenues from recorded music are set to grow. The report makes two other points that are relevant. First, that their survey of musicians that underpinned the report's data demonstrated that music creators' income from recorded music (which the survey defined as streaming, downloads and physical sales) made up only a small proportion of earnings for music creators based in the UK, with the majority of music creators whose sole income came from music-based activities supplementing their earnings from recorded music with revenue from live performance and teaching (ibid.). Second, catalogue music that is more than 12 months old now accounts for more than half of all streamed music. As we saw in the last chapter, the rise in catalogue's popularity has been meteoric; in 2015 streaming rates for new and catalogue music were equivalent (close to a 1:1 ratio according to Hesmondhalgh et al., 2021), whereas by 2019, the ratio of new to catalogue music was approximately 0.45:1 (ibid.).

The Music Creators' Earnings in the Digital Era report very much presents a picture of an economy still in the midst of a dynamic transition, but its authors are careful to make the point that, even though the size of the economy for recorded music has dropped, and that there are multiple challenges for those wanting to make a career from music, there is no evidence to suggest - even in the pre-digital heyday of the 1990s when total revenues were markedly higher than current levels – that recorded music was ever the source of 'substantial income' for significant numbers of musicians (ibid.). All of this evidence points towards a future scenario where what we think of as a traditional model for the music economy may well be utterly swept away. Ecologically speaking, the real interplay between the dramatic changes occurring in the recorded music economy and the rapidly ascendant creator economy are catalysing a variety of effects and changes in terms of how musicians create music, how they might earn a living from their music-based endeavours and how music itself might be changing. At the same time, as we have seen throughout this chapter, through the development of creator tools, and tools marketplaces, it is clear that how we categorise musicians, certainly the idea of what constitutes a 'professional' musician, is altering, as creative music tools lower barriers to entry, allowing more creators with skills that might sit outside of a traditional music skills framework, to create, share and monetise their musical works.

While a career in music as part of the creator economy may remain a real-time activity for most musicians, which is to say that royalty payments for recorded music will not play a major part in many musicians' revenue streams, generating some level of income from music and music-based activities has become attainable for more people than ever before. Even though this may not reach the stratospheric income levels of a super-rich winner-takes-all demographic of yesteryear, as Hesmondhalgh et al. (2021) point out, the truth of the matter is that there has never been a substantial number of musicians who have made a substantial amount of money from selling recorded music.

Certain key characteristics of a changing, interdependent music ecology are worthy of note. The contemporary digital environment is facilitating the production and commercial exploitation of music in two fundamentally distinct ways: rights-based revenue, and advertising-based revenue. With the growth and evolution of TikTok as a global force in social media, music, which has often been associated with social patterning, in terms of the role it has played in the formation of subcultures and as a communications medium, is currently enmeshed with a new form of digital sociality. In this context, the rise of the creator economy reflects not only the shifting economic environment in which music production, distribution and consumption take place, but perhaps more fundamentally, a shift in the relationship between creators and consumers, between professional and amateur musicians, and the role that music plays in bringing communities of music makers and listeners together. According to MIDiA, while entertainment companies have long been dominant, meaning that the professionalisation of music production, and commercialisation of distribution and consumption, creators operating in a market environment defined by different forces and trajectories, look set to dominate a future music landscape (Cirisano et al., 2021). This is to say that, not only is music changing in terms of how and where it is produced, but the long-held traditions of the music industry look set to change, and change rapidly. Although the punk revolution of the late 1960s and 1970s (from The Stooges and The MC5 through to The Sex Pistols, The Clash and The Ramones) still operated according to a centralised model of professionally curated music production and distribution, the recently emergent creator culture, which certainly draws on the early twenty-first century producer-consumer/maker-listener precedents that Gauntlett (2011) and others observed, operates according to a more decentralised, or poly-centric model. As such, while 'the creator economy' by definition associates an emergent set of creative online practices with economic productivity, we may be better served by thinking more broadly – and more ecologically – of 'the creator ecology'.

Throughout this chapter I have examined how a set of observable developments in technology, online economics and online cultures have interacted with practices of making, sharing and experiencing music. My aim has not been to suggest or emphasise that there is always a financial imperative that accompanies or informs the production of music, nor that music is something that solely exists online, and neither do I think that music is something that is largely made using digital tools. Instead, my concern in this chapter has been with examining the contemporary music landscape, in the form of the creator economy, in terms of its ecological composition, and in doing so, enable an ecological view of music itself; one that veers away from fixed characterisations of how music should function in an online environment, and what musicians should or should not be expected to do in order to make a living from music.

Moving away from theories of technological determinism, and towards an ecological perspective, we might say that the creator economy as currently understood, is not directly caused by how the Internet operates, and neither the creator economy nor the Internet has directly led to the emergent financialisation of music. Something that ecological thinking helps us to achieve – particularly in terms of Barad's theories of entanglement – is an approach that does not seek to reduce music in the online environment to being 'about' or 'the result of' anything. In the same way, it is important to avoid thinking that the changes taking place in music are fundamentally technologically determined by the Internet, or that the Internet itself is simply determined by the capitalist socio-economic environment in which it has arisen. The reading is more nuanced. If technological determination does exist, it must take its place amongst other determining factors. Theories of entanglement show us that causality is, at its very best, non-linear and non-determinable as such. Things happen, and other things happen. Things come from things.

Music, the creator ecology and the Internet are all non-linear ecological forces and environments. They are as much productions of everything else they come into contact with, of themselves, and of their moments of engagement, as they are 'of' their obvious points of origin. As agential realism suggests, a 'point of origin' is not the same as a 'product of', and it is here where the idea of a creator ecology achieves its fullest expression. Although Ek's critics might rage against his apparent devaluing of the traditional music industry, where musicians and fans alike were able to behave in a certain way, where music could be produced and paid for in a familiar and predictable way, Ek's ideas are a reminder that – as an ecology – music might happen, but there is no set process for how it should happen. In that sense, the music industry – at least the familiar model being championed by Ek's critics – does not exist. It was simply the result of a particular entanglement of forces – technological, economic, cultural, creative – that gave rise to particular forms of producing and consuming music. The music industry can thus be seen as a 'mark on a body'. In the context of 'practices of boundary making', it is a boundary where a particular arrangement of forces has over time created a mode of operating that has a particular set of characteristics, which in themselves have led to certain expectations from musicians and music fans. Thus, the creator ecology is a demonstration of the unfixed nature of the music industry. It is an ecology visibly composed of a set of interrelated forces, and these forces are now reshaping a number of aspects of creative music practice that have hitherto been taken to represent some kind of permanent essence of music. The new materialist frameworks that we have investigated enable us to see that both versions of the music economy are real, and both have real effects in terms of shaping the lives of those who are seeking to make a living from the creation of music. Similarly, these macro-frameworks are shaping how we value the experience of music, how we go about making music and, indeed, the nature of the music that is being made. Thinking in terms of the creator ecology is thus a means to remain attuned to the open-ended nature of music and of creative music practice, and to celebrate music's deep entanglement with the technological foundations of twenty-first century culture.

Notes

1 While its focus on ensuring that copyright holders receive due recompense for the use of their work has remained the same, what was implemented as Article 17, in 2019, first appeared as Article 13 in 2018 (BBC News, 2018).
2 Since 2020, on average, Spotify has continued to pay artists between $0.003 and $0.005 per stream (Ditto Music, 2023).
3 Gauntlett also put forward a longer definition, that articulated a given activity's uniqueness along with its capacity to create empathetic connections between the maker and those engaging with the activity as a viewer or listener: 'Everyday creativity refers to a process which brings together at least one active human mind, and the material or digital world, in the activity of making something. The activity has not been done in this way by this person (or these people) before. The process may arouse various emotions, such as excitement and frustration, but most especially a feeling of joy. When witnessing and appreciating the output, people may sense the presence of the maker, and recognise those feelings (Gauntlett, 2011: 76).
4 The original, publicly accessible internet launched by Tim Berners Lee in 1990 – and retrospectively branded 'Web 1.0' – was essentially a set of static webpages that displayed content retrieved from centrally-administrated web-servers. As technological progress enabled the uploading of user-generated content, which fostered a rapid rise in online social connectivity, a new paradigm emerged – Web 2.0. Google and Facebook were launched in 2004, followed by YouTube in 2005, and within three years, Gauntlett, Leadbeater and Shirky were all reflecting on the significant impact made by these platforms on mass culture.

5 The technology analyst Azeem Azhar's 2021 book *Exponential: How Accelerating Technology Is Leaving Us Behind and What to Do About It*, focused on exactly this issue; the 'exponential gap' between technological change and patterns of governance, culture and commerce (Azhar, 2021).
6 According to a report published by the UK government in 2018, the gig economy can be defined as follows: 'the gig economy involves the exchange of labour for money between individuals or companies via digital platforms that actively facilitate matching between providers and customers, on a short-term and payment by task basis' (Lepanjuuri, Wishart and Cornick, 2018).
7 For 2021, Forbes' figures showed that the creative industries generated approximately $2.25 tr in annual revenue, while the size of gig economy market stood at $300 bn, and was forecast to rise to $455 bn by 2023. Also notable was that, at the time of the report, 40 per cent of the US workforce made at least 40 per cent of their income from gig work, and 64 per cent of full-time workers said that they wanted to do 'side hustles' (Bergendorff, 2021).
8 This chapter frequently draws on the work of MIDiA Research, a market intelligence and consulting company who specialise in entertainment and digital media business analysis. MIDiA's approach to examining and articulating the evolution of music production and commerce in contemporary digital landscapes is an interdependent approach. This is to say that their reports often focus on the behaviour patterns of listeners – or in a broader sense, music consumers – across different digital platforms, changing consumption patterns, new technological opportunities, emerging revenue paradigms, along with the evolving identity and activities of music makers – now increasingly subsumed within the broad category of content creators – who are integrated in a changing creative and commercial environment that is itself rapidly evolving.
9 Micropayments and virtual goods are payment features of the creator economy that have been ported from the online gaming economy, where in-app and in-game purchases of a range of game-related artefacts enable games companies to generate significant revenue. Referencing MIDiA market data showing that 25 per cent of games video viewers and 35 per cent of live music streamers buy 'digital items' in games, Mulligan and Jopling note that by purchasing features such as digital cards, players and fans are able to accrue value for themselves against future in-app experiences, thereby demonstrating the increasingly hybrid nature of consumption and promotion (Mulligan and Jopling, 2020).
10 Lil Nas X's 2019 hit 'Old Town Road' resulted from the purchase of the amateur producer Kio's original beat from the BeatStars platform in 2018. Over 838,000 videos hosted on TikTok that have used the track (TikTok, 2022) and over 9 million views of an Old Town Road compilation video on YouTube (Blissful Mind, 2022), provide clear evidence of the parallel music economy that MIDiA discuss. It is also notable that, as of 2019, BeatStars had 1.5 million users and had paid almost $50 mn to creators using the platform (Ifeanyi, 2019).
11 According to data published by the statistics portal, Statista, TikTok had approximately 656 million users worldwide in 2021, a figure that was forecast to grow to 955.3 millioin by 2025 (Ceci, 2022).
12 While, given their size and the diverse nature of their portfolios, it might seem anachronistic to talk of Warner and Universal as 'record labels', essentially, this is what they are. Being rightsholders, they operate according to the content-first model of content producers who generate revenue based on the value and distribution of their intellectual property.
13 In its announcement of the new music revenue sharing arrangement, Meta confirmed that video creators would receive a 20 per cent share of revenue generated by eligible videos. A separate share would be apportioned to rights holders and to Meta itself (Meta, 2022).

Bibliography

Adegbuyi, F. 2021. 'Investing in influencers: 13 funds paying online creators for content'. Available at https://www.shopify.com/uk/blog/creator-fund (accessed October 2022).
Althusser, L. 1969. *For Marx*. London and New York: Verso.
Althusser, L. 1970. 'Ideology and ideological state apparatuses', in Althusser, L.2014, *On the Reproduction of Capitalism: Ideology and Ideological State Apparatuses*. London: Verso.
Azhar, A. 2021. *Exponential: How Accelerating Technology Is Leaving Us Behind and What to Do About It*. London: Random House.
BandLab Technologies. 2022. 'About BandLab'. Available at https://blog.bandlab.com/about/ (accessed October 2022).
Bartlett, J. 2015. *The Dark Net*. London: Windmill Books.
BBC News. 2018. 'European Parliament backs copyright changes'. Available at https://www.bbc.co.uk/news/technology-45495550 (accessed September 2022).
Beaumont, M. 2020. 'Spotify's Daniel Ek wants artists to pump out "content"? That's no way to make the next "OK Computer".' Available at https://www.nme.com/features/spotify-daniel-ek-three-to-four-years-controversy-criticism-2721051 (accessed September 2022).
Bergendorff, C.L. 2021. 'From the attention economy to the creator economy: A paradigm shift'. Available at https://www.forbes.com/sites/claralindhbergendorff/2021/03/12/from-the-attention-economy-to-the-creator-economy-a-paradigm-shift/?sh=2304ed75faa7 (accessed September 2022).
Blissful Mind. 2022. *Old Town Road (TikTok Compilation)*. Available at https://www.youtube.com/watch?v=LxwpKKK3P4s (accessed October 2022).
Broadhurst, R., Cirisano, T., Das, S., Kahlert, H., Langston, A., Millar, A., Mulligan, M., Mulligan, T., Severin, K. and Thakrar, K. 2022. 'Re-creating the creator economy'. Available at https://www.midiaresearch.com/reports/re-creating-the-creator-economy (accessed September 2022).
Business Wire. 2011. 'Justin.tv Launches TwitchTV, the world's largest competitive video gaming network'. Available at https://www.businesswire.com/news/home/20110606005437/en/Justin.tv-Launches-TwitchTV-World's-Largest-Competitive-Video (accessed October 2022).
Carr, A. 2021. 'Apple's 30% fee, an industry standard, is showing cracks'. Available at https://www.bloomberg.com/news/newsletters/2021-05-03/apple-s-30-fee-an-industry-standard-is-showing-cracks (accessed October 2022).
Ceci, L. 2022. 'TikTok: number of global users 2020–2025'. Available at https://www.statista.com/statistics/1327116/number-of-global-tiktok-users/ (accessed October 2022).
Cirisano, T., Mulligan, M. and Thakrar, K. 2022. 'State of the music creator economy: Post-lockdown growth'. Available at https://www.midiaresearch.com/reports/state-of-the-music-creator-economy-post-lockdown-growth (accessed October 2022).
Cirisano, T., Das, S., Kahlert, H., Mulligan, M., Mulligan, T., Severin, K. and Thakrar, K. 2021. '2022 MIDiA predictions: The year of the creator'. Available at https://www.midiaresearch.com/reports/2022-midia-predictions-the-year-of-the-creator (accessed September 2022).
Cockos Incorporated. 2022. 'NINJAM'. Available at https://www.cockos.com/ninjam/ (accessed October 2022).
Cummings, B. 2020. 'OPINION: Spotify CEO Daniel Ek's tone deaf comments reveal who the platform really works for and how it must change'. Available at https://www.godisinthetvzine.co.uk/2020/08/07/opinion-spotify-ceo-daniel-eks-tone-deaf-comments-reveal-who-the-platform-really-works-for-and-how-it-must-change/ (accessed September 2022).

Dalugdug, N. 2022. 'TikTok parent in talks with labels to launch music streaming service in 12+ new markets – report'. Available at https://www.musicbusinessworldwide.com/tiktok-parent-in-talks-with-labels-to-launch-music-streaming-servi ce-in-12-new-markets/ (accessed October 2022).

Daman Thomas, M. 2020. 'Digital performances Livestreaming music and the documentation of the creative process', in Mazierska, E., Gillon, L. and Rigg, T. (eds), *The Future of Live Music*. London: Bloomsbury Academic.

Ditto Music. 2023. 'How much does Spotify pay per stream in 2023?'. Available at https://dittomusic.com/en/blog/how-much-does-spotify-pay-per-stream/ (accessed April 2023).

Dredge, S. 2020. 'Spotify CEO talks Covid-19, artist incomes and podcasting (interview)'. Available at https://musically.com/2020/07/30/spotify-ceo-talks-covid-19-artist-incomes-a nd-podcasting-interview/ (accessed September 2022).

Elk. 2022. 'Elk LIVE is… live!' Available at https://elk.audio/elk-live-is-live/ (accessed October 2022).

European Commission. 2022. 'Copyright: Commission urges Member States to fully transpose EU copyright rules into national law'. Available at https://ec.europa.eu/commission/presscorner/detail/en/IP_22_2692 (accessed September 2022).

Fox, C. 2019. 'What is Article 13? The EU's copyright directive explained'. Available at https://www.bbc.co.uk/news/technology-47239600 (accessed September 2022).

Gauntlett, D. 2011. *Making Is Connecting: The Social Meaning of Creativity, from DIY and Knitting to YouTube and Web 2.0*. Cambridge, Malden: Polity.

Geeter, D. 2019. 'Twitch created a business around watching video games — here's how Amazon has changed the service since buying it in 2014'. Available at https://www.cnbc.com/2019/02/26/history-of-twitch-gaming-livestreaming-and-youtube.html (accessed October 2022).

Gittleson, K. 2014. 'Amazon buys video-game streaming site Twitch'. Available at http s://www.bbc.co.uk/news/technology-28930781 (accessed October 2022).

Hall, P. 2022. 'Review: Elk live bridge'. Available at https://www.wired.com/review/elk-live-bridge/ (accessed October 2022).

Hern, A. 2018. 'The rise of Patreon – the website that makes Jordan Peterson $80k a month'. Available at https://www.theguardian.com/technology/2018/may/14/patreon-rise-jordan-peterson-online-membership (accessed October 2022).

Hernandez, R.R. 2020. 'These three collaboration apps are changing the way music is created'. Available at https://www.forbes.com/sites/robonzo/2020/08/11/these-three-colla boration-apps-are-changing-the-way-music-is-created/ (accessed September 2022).

Hesmondhalgh, D., Osborne, R., Sun, H. and Barr, K. 2021. *Music Creators' Earnings in the Digital Era*. Available at https://www.gov.uk/government/publications/music-crea tors-earnings-in-the-digital-era (accessed October 2022).

House of Commons Digital, Culture, Media and Sport Committee. 2021. 'Music streaming must modernise. Is anybody listening?' Available at https://ukparliament.shorthandstories.com/music-streaming-must-modernise-DCMS-report/index.html?utm_source=committees.parliament.uk&utm_medium=referrals&utm_campaign=economics-music-streaming&utm_content=organic (accessed October 2022).

Hyman, J. 2021. 'Patreon CEO breaks down the future of the creator economy'. Available at https://uk.finance.yahoo.com/video/patreon-ceo-breaks-down-future-151032791.html?guce_referrer=aHR0cHM6Ly93d3cuZ29vZ2xlLmNvbS8&guce_referrer_sig=AQAAAC PU_-JdqXzERimWbET37XIkZHGVbrjwVOWLJNZb4pTOW8hQ6wkiO8cfbJf-h_anT Z24y9rrj5aB3JKdXFp5Z96povKJAXEDGhygFnMXg2j1O4B0cDQv5DKRVLFZHhXX 2EP-AHfzU87IuvG_H1A7u8NQ6PJ9wBqjhy0AtVzhO-JJ&guccounter=2 (accessed September 2021).

Ifeanyi, K.C. 2019. 'How a kid from the Netherlands and a startup called BeatStars led to Old Town Road'. Available at https://www.fastcompany.com/90344161/how-a-kid-from-the-netherlands-and-a-startup-called-beatstars-led-to-old-town-road (accessed October 2022).

Kahlert, H. 2022a. 'The rise of the lean-through superfan: Post-Covid trends of UGC creation and entertainment'. Available at https://midiaresearch.com/reports/the-rise-of-the-lean-through-superfan-post-covid-trends-of-ugc-creation-and-entertainment (accessed October 2022).

Kahlert, H., 2022b. 'Lean-through consumption: Communities are key'. Available at https://www.midiaresearch.com/reports/lean-through-consumption-communities-are-key (accessed October 2022).

KateBushMusic. 2022. *KateBushMusic Channel*. Available at https://www.youtube.com/channel/UC31hY7kg7CUOBpdyn9d_gYA (accessed October 2022).

Laruelle, F. 2012. *From Decision to Heresy: Experiments in Non-Standard Thought*. Bodmin and Kings Lynn: Urbanomic.

Leadbeater, C. 2008. *We-think: Mass Innovation, Not Mass Production*. London: Profile Books.

Lepanjuuri, K., Wishart, R. and Cornick, P. 2018. *The Characteristics of Those in the Gig Economy: Final Report*. Department of Business, Energy and Industrial Strategy. Available at https://assets.publishing.service.gov.uk/government/uploads/system/uploads/attachment_data/file/687553/The_characteristics_of_those_in_the_gig_economy.pdf (accessed October 2022).

Liptak, A. 2019. 'Poland has filed a complaint against the European Union's copyright directive'. Available at https://www.theverge.com/2019/5/25/18639963/poland-european-union-copyright-directive-filed-complaint-court-of-justice (accessed September 2022).

Live Nation. 2022. 'Livestreams'. Available at https://www.livenation.com/livestreams (accessed October 2022).

Mack, Z. 2019. 'How streaming affects the lengths of songs'. Available at https://www.theverge.com/2019/5/28/18642978/music-streaming-spotify-song-length-distribution-production-switched-on-pop-vergecast-interview (accessed October 2022).

Marx, K. 1904. *A Contribution to the Critique of Political Economy*. Chicago: Charles H. Kerr & Company.

Meta. 2022. 'Music revenue sharing is globally available: Giving Facebook creators a new way to earn money'. Available at https://www.facebook.com/creators/music-revenue-sharing (accessed October 2022).

Meyer, D. 2019. 'Google once cited EU's copyright reforms as a business risk. Now they're a reality'. Available at https://fortune.com/2019/03/26/eu-copyright-directive-google-news/ (accessed September 2022).

Mulligan, M. and Jopling, K. 2020. 'The rising power of UGC'. Available at https://midiaresearch.com/reports/the-rising-power-of-ugc (accessed October 2022).

Mulligan, M. and Thakrar, K. 2021. 'Creator culture: The song becomes the feed'. Available at https://www.midiaresearch.com/reports/creator-culture-the-song-becomes-the-feed (accessed September 2022).

Mulligan, M., Thakrar, K. and Cirisano, T. 2022. 'The future of music: The rise of a counterculture industry'. Available at https://www.midiaresearch.com/reports/the-future-of-music-the-rise-of-a-counterculture-industry (accessed October 2022).

Music Business Worldwide. 2022. 'Is this the real reason Epic Games acquired Bandcamp?' Available at https://www.musicbusinessworldwide.com/podcast/is-this-the-real-reason-epic-games-acquired-bandcamp/ (accessed October 2022).

Nagle, A. 2017. *Kill All Normies: Online Culture Wars from 4chan and Tumblr to Trump and the Alt-Right*. Winchester, UK and Washington, US: Zero Books.
Netflix. 2022. *Stranger Things*. Available at https://www.netflix.com/ (accessed October 2022).
Reed, T.V. 2018. *Digitized Lives: Culture, Power and Social Change in the Internet Era*. London and New York: Routledge.
Regan, T. 2021. 'From "Fortnite" to "Roblox": The best in-game concerts ever, ranked'. Available at https://www.nme.com/features/gaming-features/fortnite-roblox-best-in-game-concerts-2021-3021418 (accessed October 2022).
Reynolds, M. 2019. 'What is Article 13? The EU's divisive new copyright plan explained'. Available at https://www.wired.co.uk/article/what-is-article-13-article-11-european-directive-on-copyright-explained-meme-ban (accessed September 2022).
Richards, W. 2022. 'Rock band to release 1,000-track album of 30-second songs to protest Spotify royalty rate'. Available at https://www.nme.com/news/music/rock-band-to-release-1000-track-album-of-30-second-songs-to-protest-spotify-royalty-rate-3155423 (accessed October 2022).
Shirky, C. 2008. *Here Comes Everybody: The Power of Organizing Without Organisations*. London: Allen Lane.
Silberling, A. 2022. 'Maybe creator funds are bad'. Available at https://techcrunch.com/2022/01/25/maybe-creator-funds-are-bad/ (accessed October 2022).
Singh, M. and Lunden, I. 2020. 'Resso, ByteDance's music streaming app, officially launches in India, sans Tencent-backed Universal Music'. Available at https://techcrunch.com/2020/03/04/resso-music-india-bytedance/ (accessed October 2022).
Sisario, B. 2021. 'Can streaming pay? Musicians are pinning fresh hopes on Twitch'. Available at https://www.nytimes.com/2021/06/16/arts/music/twitch-streaming-music.html (accessed October 2022).
Stassen, M. 2022a. 'Bandcamp acquired by Fortnite maker Epic Games'. Available at https://www.musicbusinessworldwide.com/bandcamp-acquired-by-fortnite-maker-epic-games/ (accessed October 2022).
Stassen, M. 2022b. 'TikTok owns a patent for a music service'. Available at https://www.musicbusinessworldwide.com/tiktok-owns-a-patent-for-a-music-service-and-its-hiring-for-tiktok-music-staff12/ (accessed October 2022).
Stassen, M, 2022c. 'TikTok just launched its own music distribution platform, SoundOn'. Available at https://www.musicbusinessworldwide.com/tiktok-just-launched-its-own-music-distribution-platform-soundon12/ (accessed October 2022).
Stassen, M. 2022d. 'Labels asking TikTok for share of ad revenue in new deal'. Available at https://www.musicbusinessworldwide.com/labels-asking-tiktok-for-share-of-ad-revenue-in-new-deal-talks-report1/ (accessed November 2022).
TikTok. 2022. *Old Town Road: lil nas x & Billy Ray Cyrus*. Available at https://www.tiktok.com/music/Old-Town-Road-6683330941219244813?lang=en (accessed October 2022).
Toffler, A. 1980. *The Third Wave*. New York: William Morrow.
Twitch. 2022. 'Music: Your favorite artists and all the best live performances, music production, and special events'. Available at https://www.twitch.tv/directory/music (accessed October 2022).
Vincent, J. 2015. 'Eternify's 30-second loops trick Spotify into paying your favorite band'. Available at https://www.theverge.com/2015/6/23/8830029/eternify-spotify-loop-payments (accessed October 2022).
Webster, A. 2019. 'Fortnite's Marshmello concert was the game's biggest event ever'. Available at https://www.theverge.com/2019/2/21/18234980/fortnite-marshmello-concert-viewer-numbers (accessed October 2022).

Wise, J. 2022. 'How many YouTubers have 10 million subscribers in 2022?' Available at https://earthweb.com/how-many-youtubers-have-10-million-subscribers/ (accessed October 2022).

Yun Chee, F. 2021. 'Critics still unhappy as EU clarifies revamped copyright rules'. Available at https://www.reuters.com/world/europe/eu-commission-clarifies-revamped-copyright-rules-amid-criticism-2021-06-04/ (accessed September 2022).

5

MUSIC, CREATIVE LABOUR AND ARTIFICIAL CREATIVITY

In 1981, suffering from composer's block, the composer David Cope created a computer program that would suggest notes and phrases to him, enabling him to extend the piece of music he was working on. Calling the project 'Experiments in Music Intelligence' – EMI – Cope based it on the idea of 'recombinancy': recombining music into new musical patterns and phrases in the style of a particular composer (Cope, 2022), be that Bach (Cope, 1994), Mozart (Cope, 1999b), Rachmaninov (Cope, 2008) or himself. As such, EMI was a typical first-wave AI (artificial intelligence) project, drawing on a clearly defined set of rules – in this case, Western musical harmony, along with a given composer's characteristic compositional style – to offer new notes and phrases. Throughout the 1980s and into the 1990s, Cope would continue working on the project, giving a detailed account in the book *Experiments in Musical Intelligence* (Cope, 1996). His work is a useful point of entry for reflecting on the implementation of AI in music, as it highlights three trajectories within wider debates on the development of artificial intelligence technologies.

First, the EMI project was designed to assist Cope in his composition practice, and to enhance his performance (in other words, to help him to overcome his composer's block). This reflects what has come to be known as a 'centaur' (Pitts, 2020, Cassidy, 2014), a combination of human and computer intelligence that outperforms either a human or a computer by themselves.

Second, Cope's work enables us to consider what the economist Daniel Susskind (2020) has called 'task encroachment'; the threat that AI might replace humans as workers. In March 2016, in what has been seen as a landmark match, the AI research company DeepMind challenged the global Go champion Lee Se-Dol to play Go against its AI system, AlphaGo. AlphaGo beat Lee 3–0 in a five-game series (BBC News, 2016). As noted by Susskind, and similarly by the scientist

DOI: 10.4324/9781003225836-6

James Lovelock (2019), this victory for AI represented the birth of a new paradigm in thought. The match was significant, but not just because it was a demonstration of human intelligence being beaten at playing a complex game by a human-made, synthetic intelligence – although this was clearly an important achievement for the AlphaGo development team. What characterised the paradigm shift was that AlphaGo was based on a model of intelligence that was quite unlike human intelligence. Much can be taken from AlphaGo's victory. Susskind's focus was on the potential impacts of artificial intelligence on the nature and availability of work, while Lovelock reflected on the long-term development of AI in relation to the future of human life in a more general sense. By tracking and modelling the consequences of artificial intelligence on labour markets and the future of work, Susskind's approach goes beyond simple predictions for the impacts of AI and of our future relationship with this technology by drawing on histories of disruption in labour markets caused by technological innovation. Lovelock's response to the ongoing development of AI goes further still. His concept of the 'Novacene' gestures towards a not-too-distant future age of human-AI symbiosis, in which humans and machines will find a new means to balance their respective needs and capabilities (ibid.).

Third, given his use of the phrase 'music intelligence' in the title of his EMI project, Cope's work also brings into the foreground questions surrounding the nature of intelligence itself; what is it? And if we are able understand it, then can we reproduce it? And how? According to its founder, Demis Hassabis, DeepMind's mission is to 'fundamentally understand intelligence, and recreate it artificially' (Hassabis, in Revell, 2020). For the philosopher Reza Negarestani, whose book *Intelligence and Spirit* (2018) is an examination of the nature of intelligence, 'contemplating the possibility of artificial general intelligence [...] is an expression of our arrival at a new phase of critical self-consciousness' (Negarestani, 2018). This suggests that developments in AI have already brought us to a new understanding of our own relationship with intelligence, and Negarestani goes on to propose that self-consciousness, along with a social frame that is conferred on human intelligence by our use of language, are both necessary qualities of intelligence (Negarestani, 2018). Clearly, the more we discover about how intelligence works, the more complex and challenging the task becomes of understanding and defining what intelligence is; a theme we shall return to later in the chapter.

The chapter proceeds along two trajectories. It explores consequences for working practices in music, with particular emphasis on production and distribution, looking at how AI developments are complementing human activity while simultaneously replacing human workers. At the same time, AI-based projects are offering a range of solutions to support, stimulate and improve music creativity and production, where the quest to understand the nature of intelligence is generating a number of compelling perspectives and implications. Therefore, the chapter also explores how our understanding of the nature of intelligence is impacting on traditional conceptions of creativity. In this way, rather than focus on the question of whether or not AI will take away musicians' jobs, my aim is to consider the scale and type of change that AI has introduced into the world of music making and distribution.

Key Terms

Before looking more widely at the implementation of AI in music, the following is a brief glossary of key terms that underpin the chapter.

Artificial Intelligence

According to the *Oxford English Dictionary*, Artificial Intelligence (AI) is 'the capacity of computers or other machines to exhibit or simulate intelligent behaviour' (OED, 2022). IBM nuance this definition slightly, describing AI as 'the broadest term used to classify machines that mimic human intelligence', and list speech and facial recognition, decision-making and translation as examples of the human tasks that AI has been used to predict, automate and optimise (Kavlakoglu, 2020). We shall return to the question of whether AI is modelling human intelligence, or intelligence as a phenomenon in its own right, later in the chapter.

In 1956, four researchers, John McCarthy, Marvin Minsky, Nathaniel Rochester and Claude Shannon, convened the Dartmouth Summer Research Project on Artificial Intelligence, simultaneously coining the term 'artificial intelligence' so as to differentiate it from the then-contemporary fields of cybernetics and automata theory (Nilsson, 2010: 78). As set out in the project's proposal, the basis for the research was:

> the conjecture that every aspect of learning or any other feature of intelligence can in principle be so precisely described that a machine can be made to simulate it. An attempt will be made to find how to make machines use language, form abstractions and concepts, solve kinds of problems now reserved for humans, and improve themselves.
>
> *(McCarthy et al., 1955)*

Since the inception of AI at the Dartmouth Research Project in the 1950s, its development has progressed in what has been seen as a series of waves. After an initial wave of technological innovation and research into the nature of intelligence, the philosopher Nick Bostrom describes how, by the mid-1970s, due to an increase in scepticism and a decrease in funding, AI development had fallen out of fashion; a period known as the first 'AI winter' (Bostrom, 2014: 8). In 1981, the Fifth Generation Project, a Japanese public–private partnership rekindled interest and investment, leading to a thriving period of development in expert systems, vision systems, robots, software and hardware (Russell and Norvig, 2022: 24). By the late 1980s, the project – along with similar projects in the United States and Europe – had failed to deliver on its objectives, and a second 'winter' descended. As neural networks and genetic algorithms began to overcome some of the challenges faced by the rule-based, expert systems of the 1980s, the second winter came to a close in the mid-1990s, (Bostrom, 2014: 9), although Russell and Norvig suggest that, even as late as 2005, it was still 'thawing' (2022: 28). From our vantage point in the 2020s, it is possible to see that, beyond these historical developments, in 2012 and 2013, two neural network projects ushered in a period of rapid acceleration in AI development which continues to the present day. SuperVision (also known as AlexNet) was an image recognition system built in 2010 by researchers at University of Toronto – Alex Krizhevsky, Ilya Sutskever and Geoffrey E. Hinton – and entered into that year's ImageNet Large Scale Visual Recognition Challenge (ILSVRC), and again in 2012. Designed to provide researchers with image data for training large-scale object recognition models, in 2012, the ImageNet dataset contained more than 15 million labelled high-resolution images, sorted into approximately 22,000 categories (Krizhevsky et al., 2017). According to Krizhevsky and his colleagues, 'SuperVision almost halved the error rate for recognising objects in natural images and triggered an overdue paradigm shift in computer vision' (ibid.). The SuperVision project was a significant step forward, and in 2013 a team of researchers at the then-independent AI company, DeepMind, made further progressions using a convolutional neural network across a series of tests on seven Atari 2600 games. The DeepMind project created a new deep learning model for reinforcement learning, enabling the system to set a new benchmark for AI performance on six out of the seven games, and beat a human expert on three (Mnih et al., 2013). Where AlexNet/SuperVision was a demonstration of neural networks' huge capacity for information processing, DeepMind's experiments in reinforcement learning led – subsequent to their acquisition by Google/Alphabet – to the development of the AlphaGo, AlphaGo Zero and AlphaZero projects. It has been suggested that the COVID-19 crisis further accelerated the adoption of AI in a range of commercial settings (McKendrick, 2021).

Almost 70 years on from the Dartmouth researchers' foundational work, much has been achieved across the spectrum of the project's stated aims. Indeed, as John Launchbury, the former director of DARPA's Information Innovation Office, has proposed, we are currently in the 'second wave' of AI development (Launchbury, 2016). The first wave was characterised by the implementation of what Launchbury described as 'handcrafted knowledge', wherein AI systems were programmed to employ logical reasoning within a narrowly defined set of rules or problems. Second wave systems[1] are trained on large datasets and make of use statistical learning to perform classification and prediction-oriented tasks. Although current iterations of AI are based on algorithms and data rather than programs and symbolic logic, and are thus still not yet emulating intelligence per se, this is nonetheless a different operational paradigm from approaches taken in the second half of the twentieth century (Naudé, 2021). Second wave AI has been widely implemented across an increasingly diverse range of contexts, including a variety of music-related environments from music creation and production, through to auditing and other business-related activities.

Weak/Narrow AI

The terms 'weak' and 'narrow' are often used interchangeably to refer to AI that is designed to focus on carrying out a dedicated task, such as playing chess or identifying an individual in a set of photographs. Although weak AI can be applied to complete a variety of complex tasks, it can only carry out one at a time. For example, Nick Bostrom describes how the chess program Fritz only operates in the 'narrow domain' of chess (Bostrom, 2014: 26). Self-driving cars and Apple's virtual assistant Siri are more advanced examples of weak AI, but share the requirement for human input to provide training data, and to set the parameters of their learning algorithms (IBM, 2020). Again, Naudé offers a useful qualification, stating that 'AI is nothing like human intelligence' – neither in terms of its capabilities, nor in how it makes predictions or learns, which is different from how humans learn – and that therefore, all AI currently in use should be defined as narrow AI (Naudé, 2021).

Strong AI/Human-level AI/Artificial General Intelligence (AGI)

According to IBM, strong AI, or Artificial General Intelligence (AGI), 'would perform on par with another human' (Kavlakoglu, 2020). At this stage, AGI remains theoretical. In order to exist, it would need to demonstrate an intelligence equal to that of humans, which would involve it having 'a self-aware consciousness that has the ability to solve problems, learn, and plan for the future' (IBM, 2020). For Bostrom, according to artificial intelligence scientists and philosophers, such as Hans Moravec and David Chalmers, given that 'blind evolutionary processes' have produced human-level general intelligence, then it is feasible that human engineering could produce human-level AI before the end of the century (Bostrom, 2014: 28). It is worth noting that, although Bostrom's book is now almost a decade old, and there is a clear lack of consensus as to when – or even if – human-level AI can be created, the underlying logic of why it should be possible remains compelling.

Reza Negarestani defines general intelligence in terms of

> three principal attributes: necessary abilities, the intrinsic social frame of these abilities, and their quantitative integration into a generative framework through which, in addition to becoming capable of recognising itself, intelligence can inquire into and modify its conditions of realisation and enablement.
>
> *(Negarestani, 2018: 20)*

Alongside the 'necessary abilities', which would no doubt include the capacity to be self-sustaining via the implementation of a range of activities and behaviours, Negarestani's contribution offers three fundamental attributes that we shall continue to explore throughout this chapter. First, 'the intrinsic social frame' refers to a wider arc in his work, the view that – given that we are only able to think because our thoughts are mediated through language, and that language by its very nature is a shared, social construct – then intelligence must, by definition, operate as a shared, social phenomenon. Second, as we saw in the IBM definition of strong AI, Negarestani sees that AGI must be self-aware, with the capacity to recognise itself. This self-awareness leads to his third fundamental attribute; the necessary capacity for an AGI to modify and improve itself. More than simply demonstrating a capacity to learn, Negarestani's phrase 'modify its conditions of realisation and enablement' suggests that an AGI should be able to improve its capacity to learn in what appears to be a self-reinforcing manner.

Negarestani's book *Intelligence and Spirit* is an exhaustive and thoroughgoing analysis of the nature of intelligence, and he proposes that 'self-relation is the formal condition of intelligence' (2018: 31), meaning that self-awareness is a necessary condition of any form of intelligence, including human intelligence. In this, self-relation encompasses what Negarestani terms a 'concrete self-consciousness', by no means a given state, instead, it is a 'practical achievement' in the realisation of intelligence (ibid.: 30). The extent to which artificial intelligence technically satisfies this criterion is not the aim of my exploration of AI and music. However, what Negarestani's work demonstrates is that what we might think is intelligence, and indeed, what we might wish to qualify as intelligence, is by no means a straightforward procedure.

Superintelligence

Bostrom defines superintelligence as 'any intellect that greatly exceeds the cognitive performance of humans in virtually all domains of interest' (Bostrom, 2014: 26). Similarly, IBM state that 'Artificial Super Intelligence (ASI) – also known as superintelligence – would surpass a human's intelligence and ability' (Kavlakoglu, 2020).

Lovelock used the term 'cyborg' to refer to the intelligent machines that he sees as populating the oncoming 'Novacene', an age that he describes as a stratigraphical age to come that will follow the Anthropocene, where 'lifeforms will emerge that are able to reproduce and correct the errors of reproduction by intentional selection. Novacene life will then be able to modify the environment to suit its needs chemically and physically' (Lovelock, 2019: 86). For Lovelock, 'cyborg' – literally 'cybernetic organism' – describes 'an organism as self-sufficient [as humans] but made of engineered materials (ibid.: 29). Like Negarestani and Bostrom, Lovelock proposed that cyborgs will be produced by humans, rather than emerging – as humans have done – via random evolutionary processes 'from the inorganic components of the Earth' (ibid.: 85). Lovelock's use of the term 'cyborg' was intended to reflect how the intelligent beings that populate the Novacene will have evolved as humans have, via Darwinian processes. However, although humans will have set such cyborgs on their evolutionary path, Lovelock envisaged that they will have 'designed and built themselves from the artificial intelligence systems [humans] have already constructed' (ibid.: 29).

Machine Learning

Machine learning (ML) is sometimes confused with artificial intelligence, but it is in fact a sub-field of AI. So-called 'classical' – or at least, non-deep – machine learning requires human input to structure the data that the system is learning from, often differentiating data into distinct categories (Kavlakoglu, 2020).

Deep Learning

Deep Learning is itself a sub-field of machine learning, based on the concept of imitating the interconnectedness of the human brain. It does this by using neural networks that are able to identify patterns in data. Deep learning algorithms are trained on large data sets, and are strongly associated with human-level AI (IBM, 2020). The word 'deep' makes reference to the structure of neural networks, which will have at least four layers; input, weights, threshold and output (Kavlakoglu, 2020).

Neural Networks

The development of neural networks was modelled on the way that information flows in biological neurons in the human brain, and an artificial neural network will mimic this process via a set of algorithms (ibid.). In the same way that biological neurons will fire in a certain way depending on the external stimuli – different sounds, objects, patterns, colours etc. – artificial neurons are designed to respond in specified ways to varying input data. Neural networks – such AlexNet – operate by receiving input data. In the case of image recognition, the colour of a pixel in a photograph will have been translated into numerical form, which is then processed via a set of calculations. Each neuron is weighted with a threshold value, and if the result of the calculation on the input data exceeds that threshold, then it progresses to another layer of neurons for another round of calculations. This weighing simulates what happens in the brain, where neurons fire a signal if the input stimulus exceeds a particular threshold of electrical activity. As the process of calculation continues through the neural network, more is learned about the data. Eventually the network recognises statistical patterns that relate the input data to the output data (Savage, 2022). For example, if the network has been trained on a dataset containing pictures that have been labelled as containing a trumpet, and other pictures that have been labelled as not containing a trumpet, it is then able to look at new images and assign a probability rating to the likelihood of these new images containing a trumpet.

Reinforcement Learning, Supervised Learning and Unsupervised Learning

We can also talk about AI learning processes in various ways. Reinforcement learning is a widely recognised paradigm that involves a machine learning system, or agent, ascertaining which actions in a given situation will produce the highest reward, through a process of trial and error. The key principle in a reinforcement learning scenario is that it is a closed system; there is no information, instruction or guidance given by an operator, human or otherwise. According to Sutton and Barto, two of the founders of reinforcement learning, the following are the three most important distinguishing features of reinforcement learning problems: 'being closed-loop in an essential way, not having direct instructions as to what actions to take, and where the consequences of actions, including reward signals, play out over extended time periods' (Sutton and Barto, 2015). Their formulation for reinforcement learning is that it must contain 'sensation, action, and goal' (ibid.); the agent must be able to recognise the state of the environment it is operating in, understand the significance of the actions it takes in terms of their effect on the state of the environment, and it must be goal-oriented, in a way that its goal relates to the state of the environment.

Supervised learning is a process of 'learning from a training set of labeled examples provided by a knowledgeable external supervisor' (ibid.), as we saw with the example of image recognition in the case of AlexNet. In a supervised learning scenario, the examples provided to the agent consist of two pieces of information, 'a description of a situation together with a specification – the label – of the correct action the system should take to that situation, which is often to identify a category to which the situation belongs' (Sutton and Barto, 2015). The aim of supervised learning in this scenario is that the agent should then be able to extrapolate from its experiences with the training dataset, and correctly apply its learning to situations that it hasn't encountered before.

The technology writer Brian Christian notes that reinforcement learning has had a circular history. Originating in animal-learning studies in the early twentieth century, it came to prominence during the early flowering of machine learning in the 1970s and 1980s, before emerging again in animal behaviour studies as a result of the authoritative insights it produced regarding the importance of dopamine in the brain. For Christian, if intelligence is the computational process by which we achieve our objectives, then reinforcement learning is the mechanism that accomplishes this, offering 'a powerful, and perhaps even universal, definition of what intelligence is' (Christian, 2020: 151).

Unsupervised learning, while it differs from supervised learning because it has not been trained on correct examples that it can then base future decisions on, also differs from reinforcement learning. Where the reinforcement learning paradigm attempts to maximise its reward in a given scenario, unsupervised learning attempts to find a hidden structure (Sutton and Barto, 2015).

As we progress through the chapter, I shall make reference to these concepts and perspectives so as to clarify and frame the challenges and consequences of thinking about music practice in terms of developments in artificial intelligence.

AI and Music

There has been a significant increase in AI-based music applications in recent years which, at the very least, has run in parallel to the wider AI adoption noted above; offering opportunities for music creation and production, along with analytic solutions for distribution and commerce. The history of AI in music is also a history of hyperbole about the significance and impact on the future of music. For example, in 2016, DeepMind were implementing AI for audio production via 'WaveNet: generative model for raw audio', described as a 'deep generative model of raw audio waveforms [...] that is able to generate speech which mimics any human voice and which sounds more natural than the best existing Text-to-Speech systems [...] and can be used to synthesise other audio signals such as music, and present some striking samples of automatically generated piano pieces' (DeepMind, 2016). In 2017, the journalist Stuart Dredge posed the question, 'AI and music: will we be slaves to the algorithm?' His report detailed how start-ups and subsidiaries of Google (Alphabet) and Sony were developing AI systems to create and process music, such as AI Music's project to remix music on demand to match its environment, or Vochlea's software that would translate vocal recordings into drum samples (Dredge, 2017). In 2019, Scott Cohen, the soon-to-be Chief Innovation Officer at Warner Music Group, emphasised how AI was changing traditional approaches to organisation, recommendation and search on music streaming platforms, comparing AI's power to radically transform the music industry to how hip hop and MTV transformed music in the 1980s (Dredge, 2019). Cohen's perspectives were also presented in an article titled 'The Amazing Ways Artificial Intelligence Is Transforming The Music Industry' (Marr, 2019). At the time, Cohen – whose former company The Orchard had, since 2014, been providing record labels with data that would allow them to correlate the impact of publishing a Facebook status update on the stream counts of music track – was also proclaiming that 'machines can write great music, and they have no ego' (Cohen, in Lunny, 2019). Again, such developments were being reflected more widely in headlines such as, 'My Future Songwriting Career Just Got Deleted by an AI

Music Startup' (Sanchez, 2018), 'Imogen Heap: How AI Is Helping to Push Music Creativity' (BBC News, 2020) and 'How Artificial Intelligence (AI) Is Helping Musicians Unlock Their Creativity' (Marr, 2021). Whether or not AI is able to create the most natural sounding text-to-speech audio, enslave musicians to the algorithm, or transform the music industry in amazing ways, there has undoubtedly been a proliferation of activity in AI and music.

This growing awareness of AI's increased presence in music speaks of an appreciable rise in the number of companies developing AI projects for music-related activities. LANDR (2022) and iZotope (2022) have an established track record of implementing AI technologies in music production and mastering. A range of companies, including Amper (Amper Music, 2020), Boomy (Boomy Corporation, 2020), Loudly (2021) and AIVA (2022), have focused on music creation, notably exploring AI's potential to benefit music creators working in games, film and social media. Originally launched in 2014 offering AI-powered music mastering services, LANDR has evolved to offer distribution services, music sample libraries, virtual instruments (VSTs) and plug-ins for music processing. The LANDR strapline simply says, 'Create, we'll do the rest', and LANDR's offering as 'the place to create, master and sell your music' (LANDR, 2022) suggests the breadth to which AI is now able to offer services and AI-enhanced workflow solutions across the music creation, production and distribution ecosystem. Boomy and AIVA also offer licensing packages that reflect their AI's creative contribution to a new track, via a variety of payment and royalties models. Amper Music offer a similar package, with their pledge to enable anyone 'regardless of their background, expertise, or access to resources' to create music (Amper Music, 2020). The company offers a customised music creation service where clients can specify a set of requirements that a unique piece of music is then tailored to meet. Another post-second winter AI start-up, Amper launched in 2014 with the ambition to facilitate and enable musical creativity regardless of a user's own musical background. In 2017, the musician Taryn Southern used Amper to develop and co-create the song 'Break Free', following this up with an entire album in 2018, where Amper generated melodies, chords and percussion elements (Deahl, 2018). Southern wrote and performed lyrics and vocal melodies, selected the instrumentation and set the tempos for the tracks, and saw the AI as an elaborate, empowering tool for someone like herself who had little knowledge of music theory (ibid.). Popgun proclaim that 'Popgun is leading the next generation of music AI. Our tools let anyone sing, play instruments, compose songs and master audio' (Popgun, 2021). Magenta enable their users to 'Make Music and Art Using Machine Learning' (Magenta, 2022). Flow Machines, part of Sony Computer Science Labs, are 'working with creators to generate new music using cutting-edge machine learning technology' (Flow Machines and Sony Computer Science Labs, 2022) and, since 2020, Never Before Heard Sounds have been making 'cutting-edge technology expressive and useful for musicians [and building] machine learning instruments and create powerful audio tools' (Never Before Heard Sounds, 2022). For LALAL. AI, a company who specialise in separating and extracting instrumental and vocal components from audio files, the goal 'is to make music creation and mixing easier for DJs, musicians, sound producers, engineers, as well as other music professionals and creative people,' (OmniSale GMBH, 2022). MatchMySound are 'revolutionising music learning, assessment, and play' (MatchMySound, 2022) by listening to learners as they practise playing music, comparing a learner's performance with a score of the original piece, and providing tips on how to improve. In March 2022, the audio streaming platform SoundCloud acquired Musiio, an AI start-up launched in 2018. Musiio produce AI audio tools that can '"listen" to music faster than any human possibly could' (Silberling, 2022), offering services to catalogue owners which 'automate tagging, supercharge search and provide highly detailed music analysis and reports for customers around the world' (Musiio, 2023).

Beyond such commercial applications, academic research continues in the field of AI and Music. In August 2023, the AI Music Creativity (AIMC) conference series held its fourth event, (AIMC, 2023). AIMC developed out of the integration of two other networks, Musical Metacreation (MuMe) and the Computer Simulation of Music Creativity (CSMC), with a focus on 'bringing together a community working on the application of AI in music practice [...] with topics ranging from performance systems, computational creativity, machine listening, robotics, sonification and more' (ibid.). The longer-term development of the AIMC conference demonstrates the extent to which work on the progression of artificial intelligence itself, both parallels and precedes its extension into musical contexts, but the way in which artificial intelligence and music is now an increasingly discrete area of research in and of itself. However, even as this concretisation has occurred, AI and music paradigms are being used to branch out into other disciplines. For example, 'Molecular Sonification for Molecule to Music Information Transfer' is a sonification project that uses artificial intelligence to translate molecular structures into musical compositions. The project uses molecular properties to generate the key of a new piece of music, and the bonding patterns of atoms create melodies (Mahjour et al., 2022), demonstrating a fusing of information mapping with creative musical processes. Similarly, academic scholarship has turned its attention to the impact of AI and music ecosystems. Weng and Chen's article, 'Exploring the Role of Deep Learning Technology in the Sustainable Development of the Music Production Industry' (2020), is an examination of the use of AI in the Taiwanese music industry in terms of its impact on music creation processes and aesthetic quality. For the article's authors, a corollary of the increased use of AI in music creation is that music creators should emphasise the 'unique value of their art' and that the industry more widely should both 'awaken consumers' awareness of music's quality' and at the same time seek to create more welcoming environments for listeners that would facilitate deeper listening, and wider appreciation of musicians' creativity (Weng and Chen, 2020).

As this book was being completed in early 2023, a new AI application had been rapidly gaining greater and greater attention in both the wider media and in academic debate: ChatGPT. ChatGPT is a large language model (LLM) capable of generating human-like text responses to a wide range of questions and prompts. It was developed by the Microsoft-backed, AI research and development company OpenAI, and launched in November 2022. GPT, which stands for Generative Pre-trained Transformer, is a type of LLM neural network that has been trained on large amounts of text data, and which can create a probability model by analysing the relationships between words in its training set. ChatGPT was originally based on OpenAI's GPT3.5 model, which had been trained on billions of written texts available on the Internet prior to 2022. Using this model, the application can generate responses to text-based user prompts, although, given the fact that it is not connected to the Internet in real-time, 'it can occasionally produce incorrect answers' (OpenAI, 2023). OpenAI had been backed by Microsoft since 2019, who increased their investment to $10 bn in January 2023 prior to the launch in early March 2023 of ChatGPT Plus, a new subscriber-only version based on OpenAI's newest model, GPT-4, capable of interacting with the Internet (Lindrea, 2023).

In a very short space of time, the impact of ChatGPT – or at least anticipation of the extent of its impact – has been marked. Indeed, David Boyle, co-author of the book series *Prompt* (Bowman and Boyle, 2023a, 2023b), a set of guides designed to enable corporations and musicians to harness ChatGPT for brand and career growth strategies, describes the time before ChatGPT's launch (in other words, pre-November 2022), as 'the old world', suggesting that this new world we are now living in will be utterly transformed in every way by the capabilities of ChatGPT – already at version 4 at the time of writing – to disrupt the worlds of work and creativity (Boyle, in Adams, 2023). The capacity of ChatGPT to create significant impact in music is being widely explored in terms of a range of creative applications: writing lyrics, writing music, writing code that writes music, creating marketing strategies to promote music, creating contracts to license music, and many more. It may come to pass that the impact of ChatGPT and its ilk (which at the time of writing includes Jasper AI, Quillbot, DeepMind's Chinchilla and Google Bard) creates disruption the like of which is unanticipated by any of the writers and thinkers whose work I draw on in this chapter. However, it remains to be seen whether this disruption is of a different type from that articulated by the likes of Nick Bostrom, Daniel Susskind and Brian Christian, or whether it simply accelerates and intensifies the trajectories I discuss here. This being the case, I have not rewritten the chapter to include a commentary on the effects of the ChatGPT models on music, since it is unclear whether its mode of operation and its effects will be of a significantly different type from those I outline here.

Vectors in Artificial Intelligence

While the examples given above are by no means an exhaustive account of AI's use in a variety of commercial and academic music settings, this overview allows us to observe differences in how AI is being implemented in music, and some of the consequences of these implementations. The economist Daniel Susskind focuses on AI's impact on working practices and labour markets, and his work enables us to articulate how the use of AI in music stands to impact creative production and distribution. Susskind investigates a history of technological disruption in labour markets in order to explore the extent to which AI systems – the industrial machines of today – stand to displace human labour in a range of sectors, including industry, commerce and education. Susskind describes AI's capacity to out-perform, and therefore oust human labour – by offering cheaper, more convenient and, controversially, more effective services than their human counterparts – as a 'substituting force' (Susskind, 2020). Such a substituting tendency is present in how the likes of LANDR, Amper, MatchMySound and Musiio operate; offering to outsource a range of established and emergent services (the likes of mastering, music education and catalogue tagging) to AI-powered automation. At the same time, Susskind refers to technological systems that are designed to enhance, rather than replace, human performance at a given task, by offering new opportunities for innovation and employment as a 'complementing force' (ibid.). Never Before Heard Sounds, with its aim to use AI as a tool that can enhance human creativity and productivity, very much fits into this category. By invoking the concepts of substituting and complementing forces, Susskind is referencing a history of the impact of technology on the workplace which has been characterised by technology's capacity to create new roles for human workers at the same time as it has made workers redundant in established roles. Susskind's view is that artificial intelligence has destabilised this balance, and the proliferation of substituting technologies now threatens to permanently outstrip the capacity of complementing technologies to create to new forms of employment. In this respect, he references the work of the economist Wassily Leontief, particularly the article 'Technological Advance, Economic Growth, and the Distribution of Income' (1983). In that article, Leontief compared the decline of opportunities for human labour in the face of technological disruption and widespread automation to the way in which the horses were at first gradually, and then completely, replaced by the tractor in agricultural production (Leontief, 1983).

Susskind builds on Leontief's analysis to suggest that, in the context of contemporary labour markets, AI systems are increasingly replacing humans via a process he calls 'task encroachment' (Susskind, 2020). Susskind's analysis suggests that machines are not competing for humans' jobs in a straightforward manner, or simply replacing their human counterparts. Instead, he suggests that the change is occurring at the level of the individual tasks that come together to constitute a job that a human would be employed to do. He maps task encroachment across three domains of work: manual (for example, operating machinery, driving vehicles, repairing appliances), cognitive (secretarial work, bookkeeping and accounting, bank and post office clerks) and affective (customer service roles in hospitality and call centres, but also, in the context of music, affective labour can be characterised as work that is intended to create an emotional response in others), suggesting that machine labour, running on AI systems that operate in the same way as AlphaGo and AlphaZero, is now encroaching into each of these areas. Ultimately, Susskind's concern is that technological task encroachment could lead to 'structural technological unemployment', a phrase he uses to denote a future scenario where there are too few jobs available that require humans (ibid.).

While it is not yet possible to say with any certainty whether these concerns about the future will be borne out, Susskind's work offers a variety of insights into how AI is shaping a range of human activities, and we can use these insights to inform our study of music. A significant development in our understanding of the nature of work and intelligence itself has arisen from the way in which AI systems engage with certain tasks. Although we are accustomed to thinking about different types of tasks across different domains, for example manual, cognitive or interpersonal, these distinctions only hold insofar as they require 'manual, cognitive or personal capabilities when performed by human beings' (ibid.). For Susskind, this is a serious underestimation of AI's capacity to encroach into the domain of human labour, since it mistakenly assumes that the human approach to fulfilling a task is the only way to fulfil that task. What is more, he suggests that the practice of labelling tasks according to how humans would carry them out runs the risk of further entrenching our ignorance of machine capabilities, which in turn could increase friction as the nature of work and the dynamics of the labour market change.

To further articulate this point, Susskind describes a step-change in how systems like DeepMind's AlphaGo operate. Rather than focusing on how a given task is performed, the likes of AlphaGo are designed to operate pragmatically, and focus instead on how well that task is performed. In other words, it is the outcome that matters, rather than the route taken. This has led to a situation where AI systems are not being built to do things in the way that humans do but better; instead, they are built to complete a task in the most optimal way. AlphaGo demonstrated that machines can perform a range of tasks without having to replicate human behaviours. This has implications not only for the future of work, but also for our expectations around the types of choices we might expect a machine to make. Indeed, much has been made of a specific move in the 2016 match between AlphaGo and Lee Se-Dol that speaks to this sense of just how much AI systems differ from, and have the capacity to confound, our expectations about how a task might be completed, or what kind of decisions a machine might make. The move – number 37 in the third game of the match – has drawn a particular amount of focus, owing to the fact that it radically diverged from what was expected – either from a human or a computer player – and because of what it implied about AlphaGo's rationale for its choice of moves. Not only does move 37 have various articles and webpages dedicated to it,[2] but it clearly made a significant impact on those participating in the 2016 match itself. David Silver, lead researcher on the AlphaGo project, commented that, 'the professional commentators almost unanimously said that not a single human player would have chosen move 37', and after checking the program himself, the response from AlphaGo was that 'there was a 1-in-10,000 probability that move 37 would have been played by a human player' (Silver, in AlphaGo, 2020), which indicated not only how unlikely the move was, but also the extent to which it confounded even AlphaGo's creators. Its opponent, Lee Se-Dol, was similarly wrong-footed by the move, but his interpretation – the essence of which has been the cause of the various online articles and reflections – is striking:

'I thought AlphaGo was based on probability calculation and it was merely a machine. But when I saw this move, it changed my mind. Surely AlphaGo is creative. This move was really creative and beautiful' (Lee, in AlphaGo, 2020).

Lee's view that AlphaGo was working creatively is precisely what Susskind is referring to in his analysis of the difference between human and machine approaches to carrying out tasks. Lee goes on to say that move 37 made him rethink his views on AlphaGo, and ask 'what does creativity mean in Go?' (ibid.). Clearly, he felt that there was more to AlphaGo than probability calculation, and that in some way, its radically unexpected choice of move could only be accounted for in the same sense that we understand, or at least account for, acts of creativity and creative decision-making. It is also worth noting that it seems that Lee was subsequently unable to overcome the upset caused by his experiences in the 2016 match, and the 18-time world Go champion retired from the game in 2019, saying, 'there is an entity that cannot be defeated … I'm not at the top, even if I become number one' (BBC News, 2019). This suggests that – more than simply being beaten by a computer program – there was something about the encounter with nonhuman creativity that had unseated Lee. Obviously being the best at something confers a winner with a certain amount of self-satisfaction, but it also speaks of an individual's certainty that they may well have a deeper knowledge and understanding of their field than anyone else. The example of the Lee–AlphaGo tournament suggests that a different – unaccountable – form of intelligence had begun to operate, an intelligence that Lee possibly felt that he simply could not comprehend, and therefore could not compete against. For the philosopher Alexander Galloway, 'it's not simply that machines are like people, or that people are like machines, but that both entities are like something else', a something else that he likens to 'information systems' (Galloway, 2006: 106). Drawing on theories of cyberneticist Norbert Weiner, Galloway frames both humans and machines as cybernetic systems that are driven to resist entropy (ibid.). We need not extrapolate from this any further, only to say that, if – as Weiner proposes – living organisms and machines are information systems that display similar behaviours because of their shared underlying drive to stave off entropy, then artificial intelligence is not necessarily modelling human intelligence. In this sense, Lee Sedol's unsettled response to move 37 may well be due to the fact that the template for intelligence lies elsewhere, and we would be mistaken to conflate 'intelligence' with 'human intelligence'.

Susskind tracks the progression in AI from the landmark defeat of chess grandmaster Garry Kasparov by the Deep Blue system in 1997[3] through to the no less significant defeat of the champion Go player, Lee Se-Dol, by DeepMind's AlphaGo program in 2016, and AlphaGo Zero's 100–0 defeat of AlphaGo over three days in 2017. Comparing the significant differences in the number of moves that are available to players in chess and Go, Susskind relates how, in chess the player taking the first move chooses between 20 possible moves, while in Go, the first move is one of 361 possibilities. Once each player has taken a turn, for the third move on the board, the chess player has 71,852 options to decide from in terms of where they might place whichever piece they play next. For the Go player, there are around 17 billion options, and these differences become exponentially greater as the games progress. For Susskind, whereas Deep Blue harnessed the human-programmed 'brute force processing power' to lead it to victory over Kasparov, AlphaGo Zero

> had wrung itself dry of any residual role for human intelligence altogether [...] It did not need to know anything about the play of human experts. It did not need to try and mimic human intelligence at all. All it needed was the rules of the game.
>
> *(Susskind, 2020)*

Holly+ and the Centaur

In his description of the Experiments in Music Intelligence project, David Cope puts forward the view that creativity is essentially a process of creating 'subtle' and 'elegant recombinations' of existing materials, suggesting that recombinancy 'appears everywhere as a natural evolutionary and creative process' (Cope, 2022). If the wider scope of the Dartmouth summer project was to analyse and replicate human-level understanding and problem-solving capabilities, then, while this does go beyond Cope's intentions for the EMI system, Cope's system nevertheless demonstrates a number of key features that we can explore in the context of artificial intelligence and music. In Cope's view, 'all great books' written in English are formed using 'recombinations' of the 26 letters of the English alphabet, in the same way that 'most of the great works of Western art music' are recombinations of the 12 notes of the equal-tempered scale (taking into account octave variations) (Cope, 2022). In this sense, we can understand recombinancy as a supervised training process; what is now commonly known as a model. Deployed in a variety of machine learning systems, models consist of rulesets or procedures that enable a system to identify patterns in data, or to predict behaviour, and are widely used, including in healthcare, cybersecurity, business, education and smart cities, and underpin the operating systems of virtual assistants and recommendation systems (Sarker, 2022). Inasmuch as it was trained on the technical and stylistic approaches of Bach, Mozart and Rachmaninov, and programmed to achieve the specific goal of creating new music that Cope could develop into finished pieces, EMI is an example of a first wave, narrow AI. What is more, in the same way that

Kasparov used the programme Fritz 5 to play centaur chess, EMI facilitated centaur musical creativity through the alliance of a human composer and computer-based recombination of musical training data.

As we have seen, the AI start-up Never Before Heard Sounds (NBHS) use machine learning to create audio tools for musicians. In 2021, they collaborated with the musician Holly Herndon and the technology researcher Mat Dryhurst to release Holly+, a project which transformed participant-generated audio into a new piece of music that had been 'sung' in Herndon's voice. NBHS describe how models can be trained on an instrument or an ensemble to capture particular sonic characteristics, but can also be trained to 'capture playing styles and articulation which is unique to an individual, place, time or culture' (Never Before Heard Sounds, 2022). Unlike Cope's EMI model, which was able to reproduce music in the style of long-dead composers, the Holly+ project was designed to 'enable other musicians to create new music and material from the dataset, collaborating in unconventional ways and producing new mashups and hybrids' (Never Before Heard Sounds, 2022).

Trained on an extensive set of Herndon's vocal recordings, the Holly+ model combines the textural and timbral qualities of this training set with the pitches and rhythms of up to 5 minutes of user-generated audio content (uploaded to the Holly+ website), to create a new piece of audio which, in the words of NBHS themselves, 'is imperfect and idiosyncratic' and which can produce 'unexpected (and sometimes bizarre) combinations of the model's training dataset and the input performance' (ibid.). The system is based on NBHS's own custom-made machine learning model, combining speech-synthesis algorithms adapted for musical audio, and a neural net trained on vocal and instrumental recordings. NBHS are keen to point out that these recordings are either self-produced or properly licensed, and their approach to developing and deploying models is frequently punctuated with references to their commitment to recognise, acknowledge and celebrate the fact that the model training process is completely rooted in the characteristics of human beings:

> It is important to us that we tell the story of our models. They came from a specific time, place and person [...] The output is imperfect and idiosyncratic, and produces unexpected (and sometimes bizarre) combinations of the model's training dataset and the input performance.
>
> *(ibid.)*

It is therefore apparent that, while the NBHS systems are designed to use technological innovation so as to engender creative innovation – indeed their aim is to use the model to produce digital audio and to operate as a real-time processing tool for live performance – there is a clear focus to harness this technology within an ethically informed framework, one that celebrates and acknowledges the genetic, cultural, social and creative contributions of human participants. As such, where the likes of Michie and Kasparov were confident that human–computer centaurs could outperform either human or computer opponents in chess, Never Before Heard Sounds recognise the tangible – and unique – contributions that humans make to centaur creativity. It is interesting to note that music and technology journalist Cherie Hu draws a parallel between musical creativity and making moves in a chess game in her discussion of Herndon's 2019 release 'PROTO', her and Dryhurst's previous exploration of AI and musical creativity. Music made for that release was developed using a voice model that had been trained using a process known as SampleRNN (RNN standing for recurrent neural network). The system, called Spawn, created new audio samples by analysing an existing audio file and drawing on that analysis to predict what could come next (Hu, 2019). Even at this initial stage of experimentation, Herndon was clearly interested in the significance of human–computer interaction in AI systems. For Spawn, the centaur dynamic was present in the experience of using human-generated data within the training set, alongside the 'human bias' she felt was present in Spawn's programming, suggesting that while Spawn itself was a synthetic intelligence, the decisions that were made about how to design and construct it were taken by humans (Herndon, in McDermott, 2020).

In the context of our wider project about the implications of artificial intelligence for creative music practice and related ecologies of music, the Holly+ and Spawn projects provide us with a range of insights for thinking about the learning we can take from the emergence of human–computer AI centaurs. Herndon and her Holly+ collaborators Never Before Heard Sounds demonstrate an awareness of ethical and material consequences of using AI to co-create music, discussing potential impacts on musicians as creative workers, along with Herndon's views on how the grain of human presence still persists within this highly technologised domain. NBHS are clear in their commitment to creating AI tools whose purpose is to enhance, rather than replace, humans, stating, 'our approach to machine learning is not to create systems which supplant musicians, but put these models into musician's hands and make the process creative and fully attributed' (Never Before Heard Sounds, 2022). By the same light however, Herndon, with reference to the ongoing evolution of AI audio technologies in the form of DeepMind's aforementioned WaveNet project and Google's text-to-speech tool Tacotron, is of the opinion that 'generating convincing spoken and sung voices will soon become standard practice for artists and other creatives' (Herndon, 2021), and highlights the potential of AI models to create positive impacts on creative labourers' income in other ways.[4] She discusses Voice Model Rights, and the importance of attributing and compensating the individuals (such as herself) whose voices have been used to train voice models. Herndon cites established legal precedent in the cases of Bette Midler and Tom Waits, both of whom won legal cases against commercial impersonations of their voices and vocal performances.[5] NBHS also detail their position regarding this emergent form of intellectual property, stating:

> We believe machine learning affords a novel approach to music making and listening. The generative nature of these algorithms allows musicians to not only capture the output of their musical processes, but the process itself. This creates a whole new kind of intellectual property, 'the model'.
>
> *(Never Before Heard Sounds, 2022)*

Hu is similarly sanguine about the future of creative activity as remunerated labour, stating that, although artists may 'relinquish themselves as training sets for machines', they stand to be empowered, rather than diminished, by their technological collaborations, provided it is artists who are able to take a leading role in 'public conversations about AI' (Hu, 2019).

While it is encouraging to hear positive perspectives on human–machine creative and commercial relationships, as with many other thinkers and writers engaging with AI, Susskind alights on Deep Blue and the AlphaGo projects as evidence that, in the final analysis, centaurs will go extinct, in the same way that so many other fantastical species have done over the millennia. Writing in the article 'AI Futures: How Artificial Intelligence Will Change Music' (2021a) Declan McGlynn expresses his concern about reaching the point where a musical artist's or producer's stylistic idiosyncrasies – 'every detail of a lifetime of musical and technical mannerisms and characteristics' – can be reproduced by an AI model; when, for example, it becomes impossible to tell the difference between a real and a fake Beyoncé (McGlynn, 2021a). Deepfakes are undoubtedly becoming an increasingly common feature and, as Herndon, suggests, the synthetic voice – as has autotune – may yet develop into a ubiquitous and pervasive presence in music production. As with the problems caused by deepfake technology, Susskind focuses on AlphaGo Zero as evidence that the centaur approach will ultimately be superseded by machines. AlphaZero, a generalised version of AlphaGo Zero able to play Chess and Shōgi, demonstrated this in 2017, when, after spending a day training its neural networks via reinforcement learning – in other words without any human input or reference to human-based examples – it beat the 11-time winner of the Top Chess Engine Championship, Stockfish in a 100-game match, winning 28 games, and drawing the rest. In the same way that Lee Se-Dol felt that move 37 was an indication of AlphaGo's creativity, for James Lovelock, given that AlphaZero searched 80,000 positions per second during gameplay, as opposed to Stockfish's search of 70 million, the match demonstrated that, rather than relying on the brute force that Michie described, AlphaZero was using 'some form of AI intuition' (Lovelock, 2019: 80).

Whereas Kasparov expressed the view that, as with centaur chess, human-plus-machine partnerships are the 'winning formula across the entire economy' (Susskind, 2020), for Susskind, the fact that AlphaZero was unbeaten in 100 games against Stockfish demonstrates that this is no longer the case. His view is that human–machine centaurs will continue to offer optimum possible performance across a range of domains – either in chess, the economy, or indeed, music – only as long as the computer half of the centaur is not able to bring to the partnership what the human half is able to. As with his rationalisation for the relentless progression of task encroachment, where roles are incrementally eroded rather than being displaced wholesale, Susskind sees that as computers are increasingly able to replicate human performance across a number of domains, the opportunity for human contribution diminishes (ibid.). This being the case, the centaur model will ultimately collapse, as machines outstrip human performance. Where NBHS remain positive and committed to the importance of human creative labour, Herndon's views, informed by the increased prevalence of so-called deepfake voice technologies,[6] suggest that Susskind's grim predictions about the longer-term threats posed by AI to human labour are not unfounded.

Replacement Anxiety

As with rising concerns over the use of deepfake voice technology for criminal ends, it may be that the greatest impact that AI has within music is an economic one. Susskind senses that the complementing force of technology – the force that creates new opportunities for human employment as fast as technology destroys the old roles – in the long-term will itself be disrupted by its opposite, the substituting force. The music sector has witnessed the steady and continuous growth of a variety of automated services, particularly in the field of creation, production and mastering. In the field of music creation, AIVA is a neural network that – similar to David Cope's EMI – creates classical music based on large training sets of classical music libraries. Amper achieves similar ends in the world of soundtracks for games and other screen media (Price, 2022), and Loudly AI Studio – whose website proudly states, 'trained on a catalogue of 8 million music tracks and harnesses our database of 150,000 audio sounds to generate complex music compositions in under 5 seconds' (Loudly, 2021) – does the same for popular music across a wide range of genres.

A notable example of how AI is being used to provide accessible and affordable services for musicians, by substituting a technological system for a skilled human worker, is LANDR. In its original form, LANDR provided an AI-driven mastering service, a postproduction process traditionally carried out by a mastering engineer, whereby sound equalisation and compression are applied to a musical track, or set of tracks, to create a master version from which all distributed versions would be generated. As already mentioned, LANDR now offer a variety of products including distribution services and creative tools, such as sample libraries, virtual studio tools and music production plugins. In 2021, LANDR upgraded and relaunched its proprietary mastering engine as Synapse, a system trained on 19 million mastered tracks totalling over 1 million hours of music. The LANDR website states that Synapse will 'give your music instant, professional polish at a price that works for your budget' (LANDR, 2022), strongly suggesting that this once highly specialised service can – at least for the mass market – be replicated and offered more cheaply by an AI.

In discussing their products, the likes of LANDR and iZotope, another sector-leading specialist in AI-driven audio production technology, communicate carefully constructed messages that both promote the virtues of intelligent production tools, while at the same time celebrating and re-affirming the important roles that humans play in the audio production process. LANDR's Product Director emphasises that LANDR does not think of 'AI-mastering as a question of OR, but rather as an AND that assists creators when needed' (Bourget, in Price, 2022), and positions the LANDR toolset as a tool for producing feedback during the production process, rather than producing a finished product to professional standards, ready for public consumption. iZotope offer a range of AI-driven 'Assisted Audio Technology' tools, including the Mix, Vocal and Reverb Assistant, the Repair Assistant, which uses machine learning to enable audio editors to remove unwanted noise from a track or to improve audio quality, and the Master Assistant, which allows music producers to audition different mastering styles. Like LANDR, iZotope are keen to stress that, 'each Assistant is simply a tool for a producer or engineer to streamline their work, not a replacement for that producer or engineer' (Messitte, 2021), and emphasise this point by positing a distinction between creativity and production: 'one of our major goals as a company is to find solutions to eliminate time-consuming audio production tasks for our users so they can focus on their creative visions' (Nercessian, 2018). This sentiment is further emphasised by Melissa Misicka, Director of Brand Marketing at iZotope, who describes the Assisted Audio Technology tools as 'a studio assistant who can take that first pass at repairs or a mix for you while you go get a coffee', allowing a seasoned audio professional to '[get] to a starting point more quickly' (Price, 2022). Although presented as a positive message about streamlining studio workflows, it would seem that while the mastering engineer's role may not be threatened – yet – other human roles have already been replaced: the studio assistant.

Speaking in 2019, Herndon reflected on the seemingly ceaseless forward march of AI in music:

> Every trend in electronic music is subtly telling us to automate everything: 'Don't work with people in a studio, because it's too expensive. Don't bring more people on tour. Don't even write your own music...' I love automated tools like arpeggiators and pitch correction, but we sometimes have to question what's really worth automating, versus what behaviours are worth saving in the creative and performance process. If a computer takes over certain jobs I no longer have to do, that should be freeing me up to be more human with the people around me.
>
> *(Herndon, in Hu, 2019)*

The iZotope and LANDR products effectively illustrate Susskind's concept of task encroachment in action in the audio production sector. This is to say that by segmenting production workflows into creative activity and 'time-consuming' production tasks, the iZotope assistants are catalysing task encroachment in exactly the way Susskind describes: the fragmentation of holistic processes into a set of sequential tasks that can be accomplished – often more efficiently – by automated systems. The likes of iZotope and LANDR are transforming the world of audio production; not by replacing human jobs wholesale, but by altering the nature of the work involved in the production workflow that takes music creation from inception to distribution and public consumption.

In 2016, the audio recording technology magazine *Sound On Sound* ran an article canvassing the opinions that prominent mastering engineers voiced about LANDR and other automated audio services. Many felt that, for various reasons, human mastering engineers would continue to outperform their disruptive AI counterparts. Among the perspectives expressed were that human mastering engineers were able to communicate more effectively with their human musician clients, that humans were better at mastering groups of tracks in order to produce a coherent and well-balanced album, and that, similarly, human mastering engineers were able to make musical decisions about the work that they were carrying out in order to achieve creative outcomes, the likes of which were beyond the capacity of an AI-powered mastering machine. On this last point, again it is worth keeping in mind Lee Se-Dol's thoughts on the creative nature of AlphaGo's move 37, for what we can clearly see here is precisely the recurrent hubristic error that concerns Daniel Susskind. Just because we understand – or at least think we understand – human intelligence, it by no means follows that we understand what intelligence is. As LANDR co-founder Justin Evans asked, 'In the '40s, mastering was part of the job of a transferring engineer. Then vinyl and radio changed that. Now we are living in a streaming world. What should mastering be now?' (Evans, in Inglis, 2016). Sam Inglis, author of the *Sound On Sound* article, noted that not only are AI-driven production services continually evolving in their capacity to both emulate and improve on the performance of human audio production engineers, but the changes to production workflows that they offer are emblematic of broader technological evolution: 'as things stand, LANDR and CloudBounce are not competing with human mastering engineers but operating in an entirely new market, opened up by wider changes in the way music is created' (Inglis, 2016).

Responding to the question of whether she felt that using AI to create music was cheating, Taryn Southern replied that, 'if music is concretely defined as this one process that everyone must adhere to in order to get to some sort of end goal, then, yes, I'm cheating [however], the music creation process can't be so narrowly defined' (Southern, in Deahl, 2018). Such issues – questions about what mastering 'should be', what constitutes the creative process, and what will result from the evolution of the production environment in which mastering operates – are shaping the debate that AI is increasingly asking us to engage with about creativity, production and associated musical practices.

Where LANDR and iZotope are challenging accepted norms around audio production workflows, other developments in AI are pushing on familiar notions of what creating and listening to music might be. Dadabots, an AI-music project developed by music technologists CJ Carr and Zack Zukowski, have created various musical projects using modified versions of the SampleRNN architecture used for Holly Herndon's SPAWN. Carr and Zukowski describe Dadabots as 'a cross between a band, a hackathon team, and an ephemeral research lab', engineering software so as to create music and collaborate with other musicians (Dadabots, 2022). Unlike LANDR's and iZotope's commercial focus, Dadabots is a set of experiments and explorations into AI's capacity to produce music. Short-form musical projects include 'Top 4 Music Deep Fakes in the Style of Nirvana covering Gorillaz' and 'Frank Sinatra bot sings Toxic by Britney Spears Impossible cover song video'. Working over more extended timeframes, 'Human Extinction Party' was a 100-day-long livestream of 'Neural Death Metal' in the style of the band Cannibal Corpse, and 'Outerhelios' was a 'Neural Free Jazz' livestream. The Outerhelios neural network was trained by listening to saxophonist John Coltrane's *Interstellar Space* 16 times, after which it was able to produce a stream of continuous music in the style of the classic Coltrane album, which saw him playing a series of duets alongside drummer Rashid Ali. Excerpts from the Human Extinction Party and Outerhelios livestreams are now archived as 10-hour videos on YouTube (ibid.). The Relentless Doppelgänger project, described as 'Neural Technical Death Metal' is an infinitely evolving and continuous livestream of music inspired by the Canadian death metal band Archspire, which has been running since 2019. Although it appears highly unlikely that any of the Dadabots oeuvre has been designed to challenge and replace human creative artistry, Carr and Zukowski playfully acknowledge this wider context that surrounds the evolution and pervasive growth of AI: 'and in the future, if musicians lose their jobs, we're a scapegoat. jk. Please don't burn us to death' (ibid.).

As with LANDR and iZotope, these projects are a musical reframing of Susskind's challenge to think about how such systems are fracturing and fragmenting once familiar and trusted job roles and creative labour activities. Although different in kind, the extended play livestreams and the indefinitely enduring Relentless Doppelgänger ask questions similar to those posed by their more commercially focused AI relatives. Just as Dadabots' infinite stream of music, and Musiio listening to and categorising music faster than any human can comprehend, what we think music is, and what we think it is for, is readily changing. The question may not be so much, will this or that music production tool put humans out of work, but instead, will music survive AI in any way that we will still recognise it?

Music creation – possibly the alchemical zenith of AI research and development – has continued to evolve, both in terms of the approach taken to facilitating AI-enhanced creativity, as well as how the resultant tracks can be used. AIVA (2022) offers users the opportunity to compose using preset styles (including modern cinematic, tango, fantasy, sea-shanty and many more besides) or in the style of a pre-existing track. The latter reflects an approach often used by directors, where a temporary piece of already existing music is placed alongside a film sequence to convey a sense of the desired mood or atmosphere that they are trying to achieve, or even the genre of music they would like a composer to create. Shortcutting this process, AIVA's 'compose with influences' function enables visual creators to upload a template track from which the AI can generate an original piece of music that can be used without the complications of either paying for permission to use an already-existing piece of music, or having to negotiate with a composer to commission a score that meets a director's requirements. To accompany its dynamic approach to music creation, AIVA also offers users a suite of rights packages that allow for a variety of access and ownership of its AI-generated music, enabling free, but limited, access through to a monthly subscription that gives subscribers full ownership of any music they create using AIVA, and extensive (although still not complete) access to their content. Boomy, whose landing page offers users the chance to 'create original songs in seconds, even if you've never made music before' is clearly more overtly focused on servicing a market concerned with creating content for social media platforms such as TikTok and YouTube (Boomy Corporation, 2020). The website also offers visitors a glossary of musical terms, providing simple explainers for key terms such as composition, production, mixing and tempo, provides simple tutorials to enable creators to sing over the AI-generated tracks, and offers a pricing system that enables royalty payments to be shared between Boomy and content creators themselves.

In various ways, these examples of AI-powered music creation are evidence that the roles traditionally played by composers and songwriters are being eroded by AI. However, each of these scenarios suggests that there is more happening here than simple task encroachment or worker displacement. Although – as may be expected – Amper co-founder Michael Hobe was keen to stress Amper's emphasis on empowering creatives rather than replacing musicians' jobs – saying that his interest lay in enabling 'more people to be creative and [allowing creatives to] further themselves' (Hobe, in Deahl, 2018) – Southern's remark about how we might define creativity may be more to the point.

AI and Ethics: Aligning Humans and Machines

Writing in *Superintelligence* (2014), his book-length analysis of existential threat posed to humanity by artificial 'general intelligence' and the consequential need to establish powerful control systems and mechanism, Nick Bostrom asks, 'How can we get a superintelligence to do what we want? [and] what do we want the superintelligence to want?' (Bostrom, 2014: 256). Bostrom's key concern is what he sees as the likelihood of artificial intelligence – which he frames as both 'general intelligence' and 'superintelligence' – as coming to outperform human cognitive capabilities 'in virtually all domains of interest' (Bostrom, 2014: 26). His concerns about a future superintelligence essentially stem from his vision of a future where a superintelligence project establishes a 'decisive strategic advantage' over humanity (ibid.: 123). He describes such an advantage as being the point at which a superintelligence is able to form a 'singleton': 'a sufficiently internally coordinated political structure with no external opponents' (ibid.: 124). The future scenario that Bostrom imagines is therefore one in which the human race is supplanted by artificial intelligence as the dominant force on planet earth. Thus, much of Bostrom's focus is on discussing and offering design principles for the construction of suitably robust and future-proof control mechanisms. His question as to the importance of understanding what it is that we want artificial intelligences to want, therefore centres on designing goals for AI that are aligned with human goals, or at least designed in such a way that they do not come into conflict with human interest at some point the future.

Bostrom makes use of AI theorist Eliezer Yudkowsky's concept of 'coherent extrapolated volition' (CEV) to frame how it might be that we are able to program an AI with a goal-oriented, decision-making mechanism that is able to identify, understand and in some way act on a set of complex, and sometimes conflicting, desires or objectives. The book is an exhaustively detailed analysis of the dangers of unfettered superintelligence and of the myriad challenges involved in both designing and implementing suitable control systems, and in this sense, Bostrom acknowledges the complexities and contradictions of creating a CEV-based system where human desires can be mapped and acted on. In his analysis of Yudkowsky's description of the CEV model, Bostrom likens extrapolated volition to the way in which someone might have two desires that are in conflict with each other; an alcoholic whose first-order desire is for an alcoholic drink, but whose second-order desire is to not have the first-order desire (ibid.: 260). The challenge for the CEV model is therefore twofold: to define what the concept of the CEV is, and then to actually extrapolate these two volitions in a coherent manner. Again, what is it that humanity actually wants? Bostrom also highlights that sometimes-different groupings of humans, or indeed different individuals, may have different objectives and goals in life, and different moral principles, that are simply unreconcilable. His sense is that a CEV model would not always necessarily be focused on reconciling – or 'blending' – differences into a coherent, non-contradictory single volition. Instead, Bostrom suggests that 'the CEV dynamic is supposed to act only when our wishes cohere' (ibid.: 261). Thus, when confronted with irreconcilable differences, the CEV process would not act, rather than try to forge a decision that combines all available human volitions.

In *The Alignment Problem* (2020), Brian Christian also considers the development of artificial intelligence from the perspective of AI safety, focusing particularly on a growing concern within the AI development community over what he describes as a 'catastrophic divergence' between the objectives of technology systems that are 'increasingly capable of flexible, real-time decision making' and a need to ensure that autonomous systems 'capture [human] norms and values, understand what we mean or intend, and, above all, do what we want' (Christian, 2020: 12–13).

Alighting on the issue of incentivisation, Christian cites the example of an AI agent being trained to ride a bicycle using reinforcement learning. The challenge was to ensure that the system would have to keep the bicycle upright while moving towards a target destination. The researchers running the experiment awarded the AI a small reward every time it made progress towards the destination. Christian reports that, to the researchers' astonishment, the AI rode in circles around the starting point. Although they were rewarding the AI for moving towards the destination, they had not provided it with any guidance as to whether or not it should move away from its target. For the researchers, the key insight was 'that we should strive to reward states of the world, not actions of our agent' (ibid.: 169) such as the example of the bicycle covering physical distance, where 'states' represented progress towards its ultimate goal. In short, reinforcement learning became a way of shaping behaviours by focusing on where something is, not what path it had taken to get there (ibid.).

Bostrom's and Christian's work is noteworthy, as they provide valuable perspectives on Susskind's concerns over the opposing forces of complementing and replacement technologies. Indeed, the human–technology hybrid of the centaur, where some kind of equilibrium is created between human and machine capabilities, seems to be an apposite image of the type of balance that could be achieved between humans and AI via a CEV approach. Similarly, as with Bostrom's analysis of superintelligence, 'the alignment problem' is an expression of the complexities and challenges of creating artificial intelligence that remains aligned with human interests and desires.

Towards an Ecology of Musical Intelligence

Since the Dartmouth Summer Research Project, research into AI has been focused on understanding, modelling and replicating intelligence. As we have seen, Susskind's work on outcome optimisation demonstrates that in a practical sense, intelligence – human or otherwise – is not necessarily what we think it is, and that in a development period spanning more than 60 years, a variety of artificial intelligences, while they may achieve some of the same outcomes as humans – including creating music and winning games – arrive at these outcomes in very different ways from how we do. New materialist and speculative realist perspectives allow us to extend such considerations, directing our thinking towards a more complex, ecological appreciation of intelligence. In the same way that artificial intelligence is changing traditional notions of musical creativity and music-related work, we can widen our reflections on music to think about how our understanding of intelligence itself is evolving. In this light, before bringing this chapter to a conclusion, I want to briefly examine what our growing understanding of artificial intelligence can teach us about human intelligence, and extend this learning to consider the implications for musical creativity and our relationship to music.

Reza Negarestani suggests that 'the human is a hybrid and protean idea' (Negarestani, 2020: 127), which is to say that what we understand the human to be – a self-same, continuous type of entity – is in fact a composite of forces, and fundamentally unfixed in its nature. He goes on to propose that as an idea, the human is historically constructed and that 'we can never elide the distinction between the totality of the human idea – that is the totality of its history – with what we appear to ourselves to be in any specific historical moment' (ibid.). In essence, what we humans are is different from the history of what we are. Similarly, although we might be tempted to understand ourselves in terms of essential human qualities that exist in perpetuity, when we take account of – or measure – ourselves, we realise that we are in fact completely produced by our specific historical moment. For Negarestani, the challenge is to equip ourselves with 'better conceptions of the human', conceptions that are 'open-source' and 'revisable', and which enable us to understand the rootedness of our specific limitations and capabilities 'in particular moments [and in] particular contexts' (ibid: 132). Furthermore, he suggests that the human is 'historically collective and is under construction' (ibid.: 127–8). 'Historically collective' means that, while our experience of intelligence operates at an individual level, the correct approach is to see it as the expression of a collective history. The idea that the human is 'under construction' means just that: humans are not finished, we are a process-in-motion through time. Any apprehension of 'the human' is simply a reading – a measurement – of this movement at a particular moment, and in a particular context. As we saw in Barad's work, measurements cannot present an external, complete view. The human is therefore never a complete and finished entity, and neither does a measurement reveal the entire history of human-ness, resolved into one fragmented moment. Our conception of the 'human' is merely a momentary, and specific, expression of a set of histories that cohere at the point of measurement. The idea of the human is an apprehension of a process, rather than a distillation of an essence.

Beyond these liminal factors that challenge us to recalibrate our understanding of human intelligence, Negarestani provides a set of strategies for conceptualising intelligence on its own terms. Building on the importance of understanding the human in a collective context, he proposes that the mind must be understood as the product of the 'deprivatised' nature of intelligence, and that it has an 'intrinsic social frame' (ibid.: 3). This is to say that, given our capacity to think is mediated through language – and that language by its very nature is a shared, social construct – then intelligence itself must be understood as having come into being as a corollary of its interdependent, entangled relations with the world around it.

Also of value to us, in terms of our wider project of 'mattering music' – where in this chapter we are engaging with music and artificial intelligence in the context of materialist and realist philosophies – is Negarestani's contextualisation of intelligence within a Meillassouxian model of objective, non-correlational reality. Here, Negarestani proposes that the defining feature of intelligence is its capacity to be aware of itself as something that has emerged within an objectively existing world; what he refers to as 'objective intelligibility of the conditions of its realisation' (ibid.: 35). For Negarestani, self-consciousness is ultimately consciousness of an absolute that grounds objective reality, what he describes as 'a self-consciousness that has found and secured its own intelligibility in that of an unrestrained universe' (ibid.: 37), leading him to claim that 'the first contact of intelligence with the objective intelligible world is its encounter with its own underlying structure' (ibid.: 44).

As we saw with AlphaGo's disruption of accepted frames of reference for how intelligence operates, Negarestani's ideas challenge us to think more radically about what intelligence is. Notwithstanding his imperative to radically reconsider our understanding of human intelligence, it would appear that benchmarking intelligence against human capabilities and qualities is highly problematic. In establishing the grounds for developing Artificial General Intelligence (AGI), Negarestani sees that the challenge lies in fundamentally alienating ourselves from a traditional conception of the human; looking instead to what the human is beyond frequent and familiar conceptualisations (ibid.: 118). In other words, AGI and AI are part of the attempt of our ambition to understand ourselves, to quantify and classify our own intelligence, and essentially render ourselves intelligible. If we think about how human intelligence has developed as a result of it being a property of humans as physical beings, then we can begin to establish what the necessary conditions are that will allow for and validate the emergence of intelligence and the contingent characteristics of an individual subject (ibid.: 121). For Negarestani, a defining feature of an AGI is the capacity to distinguish the 'characteristics of an objective reality from characteristics of the subject's experience' (ibid.). Therefore, successfully designing and implementing an AGI involves going beyond simply creating a framework for intelligence at the level of the human. Instead, creating an AGI would need to encompass what Negarestani sees as 'theoretical problems that have long fixed physics, cognitive science and philosophy' (ibid.: 118). In this sense, we can assume that, for Negarestani, solving the challenge of creating an AGI would also mean solving a host of practical and conceptual challenges, possibly including the so-called 'hard problem' of consciousness, which asks why and how physical processes in the brain give rise to the subjective experience of the mind, questions that have long eluded scientists and scholars from across a wide array of disciplines.

Where Negarestani speaks against the idea of the human as a fixed essence, the designer and theorist Benjamin Bratton argues that there are no such things as nature and culture. Rooting his perspectives in a material engagement with the world that acknowledges chemical, biological and geological forces, such as phase change, patterning and collapse, he proposes that human 'cognition and industry [are] manifestations of a material world acting upon itself in intelligent patterns' (Bratton, 2020: 231). Bratton equates 'artificial' with 'anomalous regularity', and sees human cognition as being artificial, since its apparent self-sameness and cohesion displays the quality that 'exceeds what could normally be expected or possible without deliberate intervention' (ibid.). Thus, human intelligence is a quality that introduces itself into the world by virtue of its ability to bestow a particular order of patterning into biological and chemical forces, and in this, Bratton sees that artificial intelligence is the capacity to recognise and denote the regular patterns that anomalously arise in the material world as 'regular'. For Bratton, debates over humanity's role in precipitating climate change bring into sharp focus the artificiality of nature and of human intelligence. Rather than debating the Anthropocene in terms of its capacity to blur the boundaries between what is natural and what is anthropogenically produced, Bratton proposes that climate change challenges us to understand that the entire world is 'an exercise in interpreting artificiality' (ibid.: 232). This is to say that the traditional practice of separating the world into the 'natural' and the 'human-made', particularly as evidenced by conceptions of the 'industrial' as opposed to the 'pastoral', is a form of attack on reality (ibid.). For Bratton, the Anthropocene is an extension of certain decisions and ways of seeing and being. The Anthropocene is not an out-of-control, human-made 'geochemical explosion', instead it is the manifestation of morally driven decisions and human control systems that are based on the deep-rooted, but mistaken, separation of humans and nature, and an assumption that nature is real, perspectives that reflect Jason Moore's notion of a 'Capitalocene', and Timothy Morton's ideas about implosive holism and 'Nature' as an essentially subscendant assemblage. Essentially, Bratton is proposing that we have mistakenly assumed that the decisions and actions which led to climate change are 'natural' and that going against this way of thinking is 'unnatural', or artificial. However, according to Bratton, artificial thinking – an epistemology that recognises the artificiality of humans and nature – is exactly what is needed to plot a successful course through the Anthropocene.

In emphasising the artificial nature of human intelligence, Bratton and Negarestani enable us to think ecologically about intelligence. For Negarestani, intelligence can only be produced via a social process. Moreover, human intelligence is enmeshed with its own history, entangled in its own process of development, and always correlated with the specific details of its historical context. For Bratton, intelligence is both part of a world that autonomously produces patterns, as well as being the capacity to recognise anomalous regularities created by these patterns. Intelligence is thus part of a productive force in the world, and at the same time, it is a quality that is able to recognise this productive force, as well as its own involvement in that force. In essence, both Bratton's and Negarestani's assertions are attempts to highlight the artificial nature of human intelligence, to produce frameworks for understanding what intelligence is beyond its human confines, and to understand how such an intelligence might be schematised and reproduced. Intelligence is not a self-same thing, it is a composite of biological, chemical, historical and social forces, produced via its own capacity to recognise itself as being both part of, and separate to, the world.

These perspectives challenge us to consider the extent to which musical programs and systems such as those we have encountered in this chapter are helping us to better understand what intelligence is and what it is not, and by extension, what the implications might be for our understanding of musical creativity. Thinking about musical intelligence from an artificial-ecological perspective is to recognise that we come to music not as fixed entities who manipulate sounds in certain ways in order to produce musical outcomes. Instead understanding the artificiality of musical intelligence is an acknowledgement that we are shaped by music, just as it is shaped by us. Humans are interdependently entangled with music, and our capacity to understand ourselves as beings who can manipulate sounds to make music, and be affected by sounds when we listen to music, is part of how we define ourselves as intelligent beings. Music is part of us, and we are part of music. And yet we are separate from music. As Meillassoux and Barad show us, we cannot help but be aware that, fundamentally, we stand apart from the world, and yet the separateness of the world is itself always a part of us. What we share in common with music is the absolute real of the world that is beyond our comprehension; it is an outside, ungraspable real that is nonetheless always on the inside of our experience and knowledge. Thinking about music in terms of the artificiality of intelligence therefore enables us to understand that musical creativity is no more intrinsic than artificial intelligence itself. It is a creativity that is formed and reformed through its interrelations with the world through sound, sound-making materials, and our capacity to listen to and share sounds with each other. As music adapts and evolves, so does musical creativity. It may well be that, rather than simply displacing humans as workers, and indeed consumers, in the commercial and creative music sectors, artificial intelligence's lasting legacy will be an awareness that intelligence and creativity are matters of ecology. In this sense, an ecologies of musical intelligence and creativity are moments of apparent stability, boundaries formed through the ongoing intra-actions between humans, machines, and a world filled with sound.

Conclusions: Artificial Intelligence… Artificial Creativity

For Daniel Susskind, developments in AI will ineluctably lead to the mass displacement of human occupations. As evidenced by AI's progression through a variety of music-based settings, the signs are already here. In some ways, the Dadabots experiments are the musical epitome of this sea change; infinite streams of artificially created music, music made by robots that no human could ever listen to in its entirety. The capacity of Musiio to listen to, analyse and tag music faster than humans could, the erosion of the songwriter, composer and producer, precipitated by the likes of AIVA, Boomy, Amper and LANDR, and the usurping of human performers by deepfake and centaur technologies that – in the wake of AlphaZero – will eventually outperform and make redundant their human counterparts; all point to music beyond humans.

However, beyond Susskind's misgivings, each of the AI-music applications discussed in this chapter suggests that there is more happening here than simple task encroachment or worker displacement. While we cannot ignore his challenge to think about how such systems are fracturing and fragmenting once familiar and trusted job roles and creative labour activities across the music ecosystem, whatever we think music is, and what it is for, music itself is also changing in response to AI. The question may not be so much, will this or that music production tool put humans out of work, but instead, will music survive AI in any way that we will still recognise it?

Returning to the ecological underpinnings of this book, I will conclude with the proposal that although Susskind's task encroachment theory is compelling, we can go further and engage with music's interdependent relationship with humans, with machines, or with intelligence itself. AI is not simply atomising jobs into tasks, and replacing the task and incrementally replacing the job. AI is transforming the entire process. Just as Theodor Adorno suggested that standardisation does the listening for us (Adorno, 1941), AI will happily do the composing and the producing for us – as well as the listening. It also identifies and creates audiences, and in so doing, establishes the parameters for making music, sharing music and performing music. Although Susskind is correct to recognise that AI is fragmenting jobs into tasks, it may not be as simple as saying that job roles and tasks are being replaced and encroached upon by AI. From what we think are jobs and tasks, to what we think of as intelligence; all are only fleeting moments of measurement. The ecologies of music, machines, intelligence and creative labour are adapting and reforming. If human intelligence is no less artificial than technological intelligence, then human creativity is no more authentic, or natural than a technological creativity built from neural networks and trained using machine learning paradigms.

In the same way that Negarestani proposes that the human is a work in progress, AI brings into sharp focus the idea that music, intelligence, jobs, the economy are all works in progress. A long-held view is that technology creates new roles at the same time as it displaces human workers,[7] a view that frames jobs as we understand them now as a permanent state of affairs. To worry that AI replaces jobs is to confer a permanence on traditional notions of the labour market; to assume that, all being well, it will always exist. Similarly, to say that humans are being substituted implies that 'the human' is something that will continue to exist in perpetuity. However, humans haven't always existed, and in all likelihood, humans will not exist forever; no species on earth ever has. AI therefore focuses us on a particular moment in our development, a phase-change, a topological shift that is happening too slowly for us to be able to predict the outcome, but fast enough for us to discern that it's happening. As Bratton suggests, intelligence has always been artificial, it was just that we needed to invent artificial intelligence to enable us to see that.

If humans and machines are equally artificial and neither can lay claim to being an authentic or natural form of intelligence, then music itself is no more fixed than its human - or machine - makers or listeners. Music has never been fixed; it has always continually evolved as makers (musicians), tools (instruments, studios) and institutions (states, legal governance frameworks and corporations such as rights companies, streaming platforms and music production systems) have changed. AI may be accelerating change at the moment, but in many ways, its lasting legacy may be that it has helped us to see more clearly music's constituent parts. Artificial creativity is creativity under construction, an acknowledgement that making music has never taken one form, and cannot be distilled to an essential form or process. Just as the human is an idea under construction, music too is an unfinished idea, a process-in-motion through time.

Notes

1 According to Launchbury, where first wave AI lacked the capacity to learn, second wave AI lacks contextual awareness or the capacity to use reasoning to shape its conclusions, and he therefore proposed a third wave of AI, characterised by 'contextual adaptation'. In this scenario, AI systems create their own explanatory models of the world, employing both learning and reasoning to evaluate their own classifications of what they perceive. The concept of contextual adaptation therefore conveys how third wave systems show awareness of how their own analysis arises out of a combination of data input and rules-based reasoning, and are then able to make further adaptions based on this meta-cognition (Launchbury, 2016).
2 For example, in 2016 *Wired* magazine published the grandly titled article 'In Two Moves, AlphaGo and Lee Sedol Redefined the Future' (Metz, 2016). In addition, the website '3778.care' (named after move 37, and move 78 in the fourth game, an equally surprising move by Se-Dol), hosts an article interpreting the two moves titled 'A Board Game that Forever Changed the Relationship Man X Machine' (3778, 2022), while an IT logistics professional posted a piece on his website reflecting on the two moves, saying that the experience of watching the AlphaGo documentary made him 'reflect on and appreciate the concept of creativity' (Wong, 2021).
3 In 1972, the computer scientist Donald Michie had described how a combination of the 'brute-force capabilities' of chess programs and a human player's capacity for intuition, a knowledge of chess strategy and the ability to check the computer's proposed moves for potential flaws would result in a new type of gameplay that he called 'consultation chess' (Michie, 1972). Kasparov updated this concept, renaming it 'advanced chess', and competed using the chess program Fritz 5 in the first Advanced Chess event in 1998 (Chessgames.com, 2021a). Advanced chess, also known as 'centaur chess' and 'freestyle chess' (Cassidy, 2014; Nilsen, 2013), developed more widely in the wake of Kasparov's approach to using a computer to improve both his and its performance capabilities.
4 See Chapter 3 for a fuller discussion of how blockchain technology is being used in the Holly+ project to acknowledge and remunerate all contributors to new music made using the Holly+ system, and to curate its musical output.
5 Bette Midler was successful in suing car manufacturer Ford for impersonating her appearance in an advertisement for the Mercury Sable car, and Waits was similarly successful against Frito-Lay for their stylistic impersonation of him in a radio advertisement for Doritos chips.
6 As reported by *The Wall Street Journal* (Stupp, 2019) and Forbes (David, 2021), deepfake voice technology describes the use of AI to create vocal simulations of living humans in order to carry out fraudulent criminal activity (for example, impersonating a chief executive's voice to request a fraudulent transfer of €220,000 (Stupp, 2019)).

According to Forbes, since 2016, the Voice Conversion Challenge (VCC) has enabled an international network of institutes, organisations and researchers to compare various voice conversion systems. VCC outcomes suggest that voice conversion technology continues to increase dramatically, with features such as naturalness, speaker similarity, amount of target voice needed to create a deepfake all seeing improvement over time (David, 2021). Herndon's comments are based on her own experience of noticing a rise in the number of videos featuring AI-generated emulations of celebrities' voices.

7 A study published by Deloitte in 2015, found that, during the 144 years since 1871, technology had created more jobs than it had displaced (Stewart, De and Cole, 2015).

Bibliography

3778. 2022. 'A board game that forever changed the relationship man x machine'. Available at https://3778.care/the-name.html (accessed April 2022).

Adams, S. 2023. 'DiS005: How music can embrace the power of AI with ChatGPT expert David Boyle'. Available at: https://open.spotify.com/episode/7fGZnq5fdAc6qjWy6OvsIg (accessed March 2023).

Adorno, T. 1941. 'On popular music', in *Essays on Music*. 2002. Berkeley, Los Angeles, London: University of California Press.

AIMC. 2023. '*AIMC 2023, 30/8 – 1/9*'. Available at https://aimc2023.pubpub.org (accessed August 2023).

AIVA. 2022. 'The Artificial Intelligence composing emotional soundtrack music'. Available at https://www.aiva.ai (accessed May 2022).

AlphaGo. 2020. 'AlphaGo – The Movie'. Available at https://www.youtube.com/watch?v=WXuK6gekU1Y&t=3030s (accessed April 2022).

Amper Music. 2020. 'Music for your defining moment'. Available at https://www.ampermusic.com (accessed April 2022).

Battier, M. 2022. 'The emergence of interactive music: The vision and presence of Joel Chadabe'. *Leonardo*, 55 (1), pp. 107–111. doi:10.1162/leon_a_02180

BBC News. 2016. 'Artificial intelligence: Google's AlphaGo beats Go master Lee Se-dol'. https://www.bbc.co.uk/news/technology-35785875 (accessed April 2022).

BBC News. 2019. 'Go master quits because AI "cannot be defeated"'. Available at https://www.bbc.co.uk/news/technology-50573071 (accessed April 2022).

BBC News. 2020. 'Imogen Heap: How AI is helping to push music creativity'. Available at https://www.bbc.co.uk/news/av/technology-52236563 (accessed April 2022).

Bhattacharyya, J. 2020. 'The evolution of ImageNet for deep learning in computer vision'. Available at https://analyticsindiamag.com/imagenet-and-variants/ (accessed April 2022).

Boomy Corporation. 2020. 'Make instant music. Share it with the world'. Available at https://boomy.com (accessed May 2022).

Bostrom, N. 2014. *Superintelligence: Paths, Dangers, Strategies*Oxford: Oxford University Press.

Bowman, R. and Boyle, D. 2023a. *Prompt: A Practical Guide to Brand Growth Using ChatGPT*. Audience Strategies.

Bowman, R. and Boyle, D. 2023b. *Prompt for Artists: A Practical Guide to Career Growth for Musicians Using ChatGPT*. Audience Strategies.

Bowman, S.R. 2023. 'Eight things to know about Large Language Models'. *arXiv* doi:10.48550/arXiv.2304.00612 (accessed April 2023).

Bratton, B.H. 2020. 'The Plan', in Garayeva-Maleki, S. and Munder, H. (eds), *Potential Worlds: Planetary Memories & Eco-Fictions*. Zurich: Migros.

Cassidy, M. 2014. 'Centaur chess shows power of teaming human and machine'. Available at https://www.huffpost.com/entry/centaur-chess-shows-power_b_6383606 (accessed April 2022).

Chessgames.com. 2021a. 'Advanced chess matches'. Available at https://www.chessgames.com/perl/chesscollection?cid=1017919 (accessed April 2022).

Chessgames.com. 2021b 'AlphaZero – Stockfish (2017).' Available at https://www.chessgames.com/perl/chess.pl?tid=91944 (accessed April 2022).

Christian, B. 2020. *The Alignment Problem: How Can Artificial Intelligence Learn Human Values?*London: Atlantic Books.

Cope, D. 1992. 'Computer modeling of musical intelligence in EMI'. *Computer Music Journal*, 16(2), pp. 69–83.

Cope, D. 1994. *Bach by Design: Computer Composed Music*. Centaur Records.

Cope, D. 1996. *Experiments in Musical Intelligence*. Madison, WI: A-R Editions.

Cope, D. 1999a. 'One approach to musical intelligence'. *IEEE Intelligent Systems & Their Applications*, 14(3).

Cope, D. 1999b. *Virtual Mozart: Experiments in Musical Intelligence*. https://open.spotify.com/album/53nWZrMblfI0zEYfub8VZD?si=N2y2iz_fQei6ivgx_KdDPw (accessed May 2022).

Cope, D. 2008. *Virtual Rachmaninov: Experiments in Musical Intelligence*. Available at https://open.spotify.com/album/5pPALU3kCySKz8ygYelpG1?si=UfIEhm W3TtKz87ZTrggbVQ (accessed May 2022).

Cope, D. 2022. 'Experiments in musical intelligence'. Available at http://artsites.ucsc.edu/faculty/cope/experiments.htm (accessed April 2022).

Dadabots. 2022. *FAQ*. Available at https://dadabots.com/music.php (accessed May 2022).

David, D. 2021. 'Analyzing the rise of deepfake voice technology'. Available at https://www.forbes.com/sites/forbestechcouncil/2021/05/10/analyzing-the-rise-of-deepfake-voice-technology/?sh=5a2401a86915 (accessed May 2022).

Deahl, D. 2018. 'How AI-generated music is changing the way hits are made'. Available at https://www.theverge.com/2018/8/31/17777008/artificial-intelligence-taryn-southern-amper-music (accessed May 2022).

DeepMind. 2016. 'WaveNet: A generative model for raw audio'. Available at https://www.deepmind.com/blog/wavenet-a-generative-model-for-raw-audio (accessed April 2022).

Dredge, S. 2017. 'AI and music: Will we be slaves to the algorithm?' Available at https://www.theguardian.com/technology/2017/aug/06/artificial-intelligence-and-will-we-be-slaves-to-the-algorithm (accessed April 2022).

Dredge, S. 2019. 'Scott Cohen: "Every 10 years something kills the music industry"'. Available at https://musically.com/2019/01/29/scott-cohen-every-10-years-something-kills-the-music-industry/ (accessed April 2022).

Flow Machines and Sony Computer Science Labs. 2022. 'Augmenting creativity with AI'. Available at http://www.flow-machines.com (accessed April 2022).

Galloway, A. 2010. 'French theory: Today an introduction to possible futures. A pamphlet series documenting the weeklong seminar by Alexander R. Galloway at the Public School New York in 2010'. Available at http://cultureandcommunication.org/galloway/FTT/French-Theory-Today.pdf (accessed June 2022).

Galloway, A.R. 2006. *Protocol: How Control Exists after Decentralization*. Cambridge, MA: MIT (Leonardo).

Herndon, H. 2021. 'Holly+'. Available at https://holly.mirror.xyz/54ds2IiOnvthjGFkokFCoaI4EabytH9xjAYy1irHy94 (accessed May 2022).

Hu, C. 2019. 'Spitting image: Meet the electronic artists collaborating with their artificial twins'. Available at https://ra.co/features/3463 (accessed May 2022).

IBM. 2020. 'Strong AI'. Available at https://www.ibm.com/cloud/learn/strong-ai (accessed April 2022).

Inglis, S. 2016. 'LANDR, CloudBounce & the future of mastering'. Available at https://www.soundonsound.com/techniques/landr-cloudbounce-future-mastering (accessed May 2022).

iZotope. 2022. 'iZotope Products.' Available at https://www.izotope.com/en/products.html (accessed April 2022).

Kavlakoglu, E. 2020. 'AI vs. machine learning vs. deep learning vs. neural networks: What's the difference?' Available at https://www.ibm.com/cloud/blog/ai-vs-machine-learning-vs-deep-learning-vs-neural-networks (accessed April 2022).

Krizhevsky, A., Sutskever, I. and Hinton, G.E. 2017. 'ImageNet classification with deep convolutional neural networks'. *Communications of the ACM*, 60(6), pp. 84–90. doi:10.1145/3065386

LANDR. 2022. 'Instant mastering. Professional results'. Available at https://www.landr.com/en/online-audio-mastering/ (accessed May 2022).

Launchbury, J. 2016. 'DARPA perspective on AI'. Available at https://www.darpa.mil/about-us/darpa-perspective-on-ai (accessed May 2022).

Leontief, W. 1983. 'Technological advance, economic growth, and the distribution of income'. *Population and Development Review*, 9(3), 403–410. doi:10.2307/1973315

Lindrea, B. 2023. 'ChatGPT can now access the internet with new OpenAI plugins'. Available at https://cointelegraph.com/news/chatgpt-can-now-access-the-internet-with-new-openai-plugins (accessed March 2023).

Loudly. 2021. 'Create awesome music in seconds: Studio-quality music, the easy way'. Available at https://www.loudly.com/aimusicstudio (accessed May 2022).

Lovelock, J. 2019. *Novacene: The Coming Age of Hyperintelligence.* London: Penguin Books.

Lunny, O. 2019. '"You're probably going to be replaced": digital music pioneer issues a stark warning for the industry'. Available at https://www.forbes.com/sites/oisinlunny/2019/01/28/youre-probably-going-to-be-replaced-digital-music-pioneer-issues-a-stark-warning-for-the-industry/?sh=1a91933a7b58 (accessed April 2022).

Magenta. 2022. 'Make music and art using machine learning'. Available at https://magenta.tensorflow.org (accessed April 2022).

Mahjour, B., Bench, J., Zhang, R., Frazier, J. and Cernak, T. 2022. 'Molecular sonification for molecule to music information transfer'. *ChemRxiv*. doi:10.26434/chemrxiv-2022-g7xkl

Marr, B. 2019. 'The amazing ways artificial intelligence is transforming the music industry'. Available at https://www.forbes.com/sites/bernardmarr/2019/07/05/the-amazing-ways-artificial-intelligence-is-transforming-the-music-industry/?sh=ea89a5507212 (accessed April 2022).

Marr, B. 2021. 'How artificial intelligence (AI) is helping musicians unlock their creativity'. Available at https://www.forbes.com/sites/bernardmarr/2021/05/14/how-artificial-intelligence-ai-is-helping-musicians-unlock-their-creativity/?sh=1c2d83057004 (accessed April 2022).

MatchMySound. 2022. 'Welcome to MatchMySound'. Available at https://matchmysound.com (accessed April 2022).

McCarthy, J., Minsky, M., Rochester, N. and Shannon, C. 1955. *A Proposal for the Dartmouth Summer Research Project on Artificial Intelligence.* Available at http://jmc.stanford.edu/articles/dartmouth/dartmouth.pdf (accessed April 2022).

McDermott, E. 2020. 'Holly Herndon on her AI baby, reanimating Tupac, and extracting voices'. Available at https://www.artnews.com/art-in-america/interviews/holly-herndon-emily-mcdermott-spawn-ai-1202674301/ (accessed April 2022).

McGlynn, D. 2021a. 'AI futures: How artificial intelligence will change music'. Available at https://djmag.com/longreads/ai-futures-how-artificial-intelligence-will-change-music (accessed May 2022).

McGlynn, D. 2021b. 'AI futures: How artificial intelligence will shape music production'. Available at https://djmag.com/longreads/ai-futures-how-artificial-intelligence-will-shape-music-production (accessed May 2022).

McGlynn, D. 2021c. 'AI futures: How artificial intelligence is infiltrating the DJ booth'. Available at https://djmag.com/longreads/ai-futures-how-artificial-intelligence-infiltrating-dj-booth (accessed May 2022).

McKendrick, J. 2021. 'AI adoption skyrocketed over the last 18 months'. Available at https://hbr.org/2021/09/ai-adoption-skyrocketed-over-the-last-18-months (accessed May 2022).

Messitte, N. 2021. 'Meet the iZotope assistants'. Available at https://www.izotope.com/en/learn/meet-the-izotope-assistants.html (accessed May 2022).

Metz, C. 2016. 'In two moves, AlphaGo and Lee Sedol redefined the future'. Available at https://www.wired.com/2016/03/two-moves-alphago-lee-sedol-redefined-future/ (accessed April 2022).

Michie, D. 1972. 'Programmer's gambit'. *New Scientist*, pp. 329–332. Available at https://stacks.stanford.edu/file/druid:fk070vw3202/fk070vw3202.pdf (accessed April 2022).

Minsker, E. 2019. 'Holly Herndon weighs in on Grimes and Zola Jesus' debate about AI and the future of music'. Available at https://pitchfork.com/news/holly-herndon-weighs-in-on-grimes-and-zola-jesus-debate-about-ai-and-the-future-of-music/ (accessed April 2022).

Minsker, E. 2021. 'Holly Herndon's AI deepfake "twin" Holly+ transforms any song into a Holly Herndon song'. Available at https://pitchfork.com/news/holly-herndons-ai-deepfake-twin-holly-transforms-any-song-into-a-holly-herndon-song/ (accessed April 2022).

Mnih, V., Kavukcuoglu, K., Silver, D., Graves, A., Antonoglou, I., Wierstra, D. and Riedmiller, M., 2013. 'Playing Atari with deep reinforcement learning'. NIPS Deep Learning Workshop 2013. *arXiv* preprint arXiv:1312.5602. doi:10.48550/arXiv.1312.5602

Musiio. 2023. 'Musiio by Soundcloud'. Available at https://musiio.com (accessed May 2023).

National Cyber Security Centre. 2023. 'ChatGPT and large language models: what's the risk?' Available at https://www.ncsc.gov.uk/blog-post/chatgpt-and-large-language-models-whats-the-risk (accessed April 2023).

Naudé, W. 2021. 'Artificial intelligence: Neither Utopian nor apocalyptic impacts soon'. *Economics of Innovation and New Technology*, 30 (1), pp. 1–23, doi:10.1080/10438599.2020.1839173

Negarestani, R. 2018. *Intelligence and Spirit*. Falmouth: Urbanomic Media.

Negarestani, R. 2020. 'On an impending eternal turmoil in human thought', in Garayeva-Maleki, S. and Munder, H. (eds), *Potential Worlds: Planetary Memories & Eco-Fictions*. Zurich: Migros Museum für Gegenwartskunst, YARAT Contemporary Art Space and Scheidegger & Speiss.

Nercessian, S. 2018. 'iZotope and assistive audio technology'. Available at https://www.izotope.com/en/learn/izotope-and-assistive-audio-technology.html (accessed May 2022).

Never Before Heard Sounds. 2022. 'Never Before Heard Sounds'. Available at https://heardsounds.com (accessed April 2022).

Nilsen, M. 2013. 'Humans are on the verge of losing one of their last big advantages over computers'. Available at https://www.businessinsider.com/computers-beating-humans-at-advanced-chess-2013-11?r=US&IR=T (accessed April 2022).
Nilsson, N. 2010. *The Quest for Artificial Intelligence*. New York: Cambridge University Press.
OmniSale GMBH. 2022. 'About us: LALAL.AI'. Available at https://www.lalal.ai/about/ (accessed April 2022).
OpenAI. 2023. 'What is ChatGPT?' Available at https://help.openai.com/en/articles/6783457-what-is-chatgpt (accessed April 2023).
Oxford English Dictionary (OED). 2022. 'Artificial Intelligence'. Available at https://www.oed.com/viewdictionaryentry/Entry/271625 (accessed April 2022).
Pitts, P.J. 2020. 'Regulatory centaurs'. *Nature Biotechnology*, 38, 788–789, doi:10.1038/s41587-020-0589-x
Popgun. 2021. 'About'. Available at https://www.wearepopgun.com/x/index.html (accessed April 2023).
Price, A. 2022. 'Artificial intelligence in music production – friend or foe?' Available at https://audiomediainternational.com/artificial-intelligence-in-music/ (accessed May 2022).
Revell, T. 2020. 'Demis Hassabis interview: Our AI will unlock secrets of how life works'. *New Scientist*, 30 December 2020. Available at: https://www.newscientist.com/article/mg24833140-700-demis-hassabis-interview-our-ai-will-unlock-secrets-of-how-life-works/ (accessed May 2023).
Russell, S.J. and Norvig, P. 2022. *Artificial Intelligence: A Modern Approach*, 4th edn. Harlow: Pearson Education.
Sanchez, D. 2018. 'Also, 'my future songwriting career just got deleted by an AI music startup'. Available at https://www.digitalmusicnews.com/2018/03/22/amper-music-ai-composers/ (accessed May 2022).
Sarker, I.H. 2022. 'AI-based modeling: Techniques, applications and research issues towards automation, intelligent and smart systems'. *SN Computer Science*, 3, p. 158. doi:10.1007/s42979-022-01043-x
Savage, N. 2022. 'Breaking into the black box of artificial intelligence'. *Nature*, doi:10.1038/d41586-022-00858-1
Short, A. 2021. '13 AI tools that are at the cutting edge of music production'. Available at https://happymag.tv/ai-music-production-tools/ (accessed May 2022).
Silberling, A. 2022. 'SoundCloud acquires Musiio, an AI music curator, to improve discovery'. Available at https://techcrunch.com/2022/05/03/soundcloud-acquires-musiio-ai-music-discovery/ (accessed May 2022).
Stewart, I., De, D. and Cole, D. 2015. 'Technology and people: The great job-creating machine'. Available at https://www2.deloitte.com/uk/en/pages/finance/articles/technology-and-people.html (accessed November 2022).
Stupp, C. 2019. 'Fraudsters used AI to mimic CEO's voice in unusual cybercrime case'. Available at https://www.wsj.com/articles/fraudsters-use-ai-to-mimic-ceos-voice-in-unusual-cybercrime-case-11567157402 (accessed May 2022).
Susskind, D. 2020. *A World Without Work: Technology, Automation and How We Should Respond*. London: Allen Lane.
Sutton, R.S. and Barto, A.G. 2015. *Reinforcement Learning: An Introduction*. Cambridge, MA and London: MIT Press. Available at https://web.stanford.edu/class/psych209/Readings/SuttonBartoIPRLBook2ndEd.pdf (accessed April 2022).
Weng, S-S. and Chen, H-C. 2020. 'Exploring the role of deep learning technology in the sustainable development of the music production industry'. *Sustainability*, 12 (2), p. 625, doi:10.3390/su12020625

Wong, J. 2021. 'Move 78'. Available at https://thewonger.com/move-78 (accessed April 2022).

Zukowski, Z. and Carr, C.J. 2017. *Generating Black Metal and Math Rock: Beyond Bach, Beethoven, and Beatles*. Available at http://dadabots.com/nips2017/generating-black-metal-and-math-rock.pdf (accessed May 2022).

CONCLUSION

Quantum Ecologies of Creative Music Practice

Mattering Music: An Ecology of Music

In this book, I have set out to develop an ecological understanding of a range of creative music practices. Within that overarching frame, across the various chapters in *Ecologies of Creative Music Practice*, my aim has been to explore music in a number of contexts where it is possible to observe, and to hear, change and transformation occurring, sometimes with notable rapidity. By engaging with change, it has been my ambition to reveal the unfixed nature of music, to understand its entanglements, and to think about it as an ecology of forces. There is clearly more to music than I have been able to capture in one book, but I hope to have provided an impetus for those with enquiring ears to listen, and to hear the interdependence of music for themselves, and discover their own ecologies of creative music practice. In the Introduction, I defined creative music practice in terms of creative practices of making and listening to music, and creative practices of constructing contexts for music production, performance, monetisation, sharing and listening. What we have seen across these surveys of creative music practice is that music is always ecological. Whether it is a composite of dynamic forces, or a component part of other ecosystems – technological, economic, social or environmental – music is made of many things, and it is part of many things.

Networks, Assemblages and Continua

Music's interdependent relationships with the world around it means that it is always changing. Just as systems thinkers like Donella Meadows show how feedback loops of energy flowing into and out of systems create change,

DOI: 10.4324/9781003225836-7

thinking in terms of the flows between music and a range of earth systems, human-made environments and conceptual frameworks is central to recognising music's relational, ecological nature. In the preceding chapters, we have engaged with music in the context of its relations with technological and economic environments, and considered how practices of making and listening to music reveal the extent to which music itself is nothing but a constantly changing fusion of entangled discourses and material relations. As we begin to conclude this ecological investigation of music, I want to now turn to scientific, anthropological and archaeological perspectives, to bring sharper focus to the sense in which music, as with other aspects of life on earth, in a profound sense *is* its relations with the rest of the world.

The biologist Merlin Sheldrake suggests that studying relationships can be confusing due to their ambiguous nature (Sheldrake, 2020: 20). He notes how leafcutter ants dedicate their lives to cultivating a fungus which they feed with leaf fragments, but, given the ability of fungal lifeforms to exert influence and even control over animal life, he asks 'have leafcutter ants domesticated the fungus they depend on, or has the fungus domesticated the ants?' (ibid.).[1] Sheldrake also notes that his study and research into tropical microbes had led him to the conclusion that talking about the existence of individuals as self-same, self-contained beings no longer made sense. As someone who had begun his scientific career studying plant science, he had come to the conclusion that 'biology – the study of living organisms – had transformed into ecology – the study of the relationships between living organisms' (ibid.: 19). Given Sheldrake's comprehensive analysis of the relationships between fungi and a vast array of other lifeforms throughout earth's history, this is no mere analogy. His research demonstrates that, at the level of living systems, the world is fundamentally constituted by relations, both between and within systems. For Sheldrake, this is seen in the more overtly symbiotic relations between fungi and certain species of ant, as well as in the composite organism lichen, which he refers to as a 'mycobiont' (ibid.: 81). Lichen are composed of algae or bacteria living in the hyphae of fungi. Hyphae are the tubular structures that fungi form which themselves become entangled into mycelial networks, and through which can flow water, nutrients and electricity, which the anthropologist Anna Lowenhaupt Tsing describes as 'thread-like filaments' that spread out and tangle in the soil (Lowenhaupt Tsing, 2015: 137). Sheldrake describes how plant roots and hyphae interact as 'mycorrhizal relationships' (literally, fungus–root relationships), noting that 'connections between hyphae and roots are dynamic, formed and re-formed as root tips and fungal hyphae get old and die. These are relationships that ceaselessly remodel themselves' (ibid.: 42). As a musician himself, Sheldrake likens these mycorrhizal interactions to the way in which a group of jazz musicians improvise during a performance, 'listening, interacting, responding to one another in real time' (ibid.: 42).

In *The Mushroom at the End of the World* (2015), Lowenhaupt Tsing examines the global movements of the matsutake mushroom, interrogating among other things, the informal foraging and trade networks that result from its being a highly sought-after delicacy. Lowenhaupt Tsing's focus is on the myriad assemblages – the socio-economic supply chains and cultural traditions – that matsutake mushrooms become involved in, from their origins in red pine forests to the plates of global matsutake connoisseurs. She suggests that ecologists make use of the word 'assemblage' 'to get around the sometimes fixed and bounded connotations of ecological "community"', since it allows for a more fluid articulation of how the species 'in a species assemblage influence each other – if at all' (ibid: 22–3). Lowenhaupt Tsing sees that, in some instances, species are in competition with each other, and they may even eat each other. At other times, species may work together, and sometimes '[they] just happen to find themselves in the same place' (ibid.). Her point is that, essentially, assemblages are 'open-ended gatherings [that] allow us to talk about communal effects without assuming them' (ibid.). As we have seen with Karen Barad's work on entanglement, for Lowenhaupt Tsing, assemblages are not sites of determination, instead they are points, or zones of interaction; interactions that are not pre-determined, and neither do they always have the same outcome. This is a compelling narrative for thinking about how musicians, streaming platforms, social media platforms, artificial intelligence, rights holders, music fans, casual listeners, amateur and professional creators all come together within a non-deterministic ecology.

For Lowenhaupt Tsing, things are neither determined by their membership of an assemblage, and neither is the assemblage determined by their presence in it. Some things are just there. In this context, Lowenhaupt Tsing engages with two musical forms as a means to understand the interconnected make-up of our contemporary industrial, agricultural and commercial environments; the 'polyphonic assemblage', and what she calls the 'progress rhythm' (ibid.: 24). Her idea is that the polyphonic structuring of fugues and madrigals, 'in which autonomous melodies intertwine' (ibid.: 23), have been superseded by the apparent progress of the drive towards unification that can be heard in the classical forms that displaced the baroque music of madrigals and fugues. Lowenhaupt Tsing hears this drive towards unity in the central role that standardised rhythm patterns and metric beats have played in popular music forms of the twentieth and twenty-first centuries. Her point of reference is the way in which rhythms that drew on the rock 'n' roll paradigm underpinned the development of popular music from the 1950s onwards (ibid.). While Lowenhaupt Tsing's ideas about the progress rhythm are compelling, and she is not necessarily wrong to identify a certain standardisation of rhythm in a variety of post-1950s popular music forms, her analysis of rock 'n' roll-based musics has its limits. It should be said that there are a variety of ways of perceiving the open-endedness of popular music beyond its adherence to a metric beat; not least the sonic experiments of popular music production and the melodic openness of popular music's adoption of folk and vernacular styles such as the blues.

Lowenhaupt Tsing contrasts the tendency that she hears towards unification of rhythm and harmony in rock, pop and indeed certain aspects of classical music with the polyphonic organising principles that underpin madrigals. Hearing madrigals and fugues as polyphonic assemblages where 'one must listen both to the separate melody lines and their coming together in unexpected moments of harmony or dissonance, [and where] the polyphony of the assemblage shifts as conditions change' (ibid.: 158), she understands how these polyphonic forms can be listened to at the level of the movements of the individual voices and melodic lines. At the same time, she hears how 'sporadic but consequential coordinations' of different lines can come together, to create a variety of harmonic directions within a piece. Despite her approach to listening to the development of rock and pop music largely in terms of a certain aspect of rhythmic development, what is interesting about Lowenhaupt Tsing's perspective is that she is clearly moved by what she calls 'a revelation in listening', describing how listening to madrigals forced her to 'pick out separate, simultaneous melodies and to listen for the moments of harmony and dissonance they created together' in polyphonic music (ibid.: 24). For Lowenhaupt Tsing, this kind of listening, which involves accepting points of divergence as much as confluence, is necessary in order 'to appreciate the multiple temporal rhythms and trajectories of the assemblage' (ibid.). Lowenhaupt Tsing's ideas and instincts are thus a reminder that even the seemingly straightforward act of listening to music can in fact be a practice of listening ecologically, of listening to a subscendant whole that is always less than the sum of its simultaneously confluent and divergent parts.

In *Blocks of Consciousness and the Unbroken Continuum* (2005) a survey of the state of improvised music-making in 2005, the book's authors, Brian Marley and Mark Wastell, reference the improvising guitarist Derek Bailey's sense that 'his improvising was continuous, broken only by the moments when he set down his guitar' as a key inspiration for the book's title, stating that 'music as a continuum which musicians dip in and out of was simply too good an idea to pass up' (Marley and Wastell, 2005: 6).[2] To suggest that an individual's approach to improvising is the sonic realisation of sustained enquiry and experimentation is indeed a compelling idea, and can work very simply to provide an intuitive and attractive means of understanding a musician's ongoing improvisations. The notion of a continuum as an ecology of performance, of techniques, remembered improvisations, along with our other knowledge of music is attractive, since all combine to produce an ecology that encompasses all the improvisations we have created. It also encompasses Lowenhaupt Tsing's conception of an assemblage – albeit an assemblage that operates over an extended period of time – wherein the musical components may be in any number of relations with each other, collaborative, competitive, or simply just there.

Similarly, in *The Musical Human* (2021) Michael Spitzer approaches humanity's relationship with music over time. He examines the history of music, considering its relationship with human evolution and development, and its relationship with contemporary human life; how we listen to it, how we consume it, and how we make use of it. In his discussions of music's developmental history, Spitzer notes that the development of music has tracked the development and ongoing complexification of human culture and society. Spitzer sees that the human species, homo sapiens, has not evolved in any significant sense since approximately 40,000 BCE, a date that coincides with marked increases in the production of art (Spitzer, 2021: 10).[3] Given these ancient connections between practices of making, and humans as the sole surviving hominin species on the planet, it is Spitzer's contention that creativity is an innate part of human cognition. We might think of this as an argument about the interdependent nature of humans and music, in the sense that homo sapiens and creative practices of art and music making emerged during the same period in history. To be human is to be musical. By mapping a long-term history of musical progress to a timeline of human development, Spitzer engages with the history of music and language, reflecting Lowenhaupt Tsing's perspectives on the combinatorial aspects of polyphonic assemblages. Spitzer sees that fundamental to musical composition as well as human speech is the capacity 'to combine units into a limitless variety of new sentences or pieces' (ibid.: 29). What is more, music and language are adaptable: 'even if they originate in specific contexts, they can be repeated and ritualised – thus abstracted from those contexts' (ibid.). In this regard, neither music nor language has a pre-set purpose as such; each can be reformed and recontextualised based on its interactions with new environments and new phenomena.

According to Spitzer's narrative, humanity's innate musicality manifests as our sustained relationship to music across millennia. The discovery of bone flute shards near cave paintings situated in the most acoustically resonant areas of cave systems indicates that, even at relatively early points in our species' development, sound was being used in a way that harnessed its aesthetic qualities. The resonant properties of these spaces would have amplified a flute's volume, and the capacity of resonant chambers to create immersive sonic environments is still being harnessed 40,000 years later, as we can hear in the music that Bendik Giske recorded inside the Vigeland Mausoleum. The co-presence of musical instruments and cave paintings also suggests that music – just as we have found throughout this book – has from the earliest performances been ecological in nature; always part of something, always shaping, and being shaped by its relation to other practices. Music, humans and art. Alluring though these histories may be, we will probably never know what was happening in the cave systems that Spitzer discusses. Were these early humans using art and music to tell stories? To engage in ritualistic experiences? Or, as we saw in the Introduction, just like the nightclubbers and ravers that Gilbert and Pearson give voice to in their analysis of music and club cultures, has our species – from the outset – simply enjoyed dancing to music as a way of giving expression to the visceral sensation of sound moving through our bodies? Whichever of these scenarios we find appealing, the idea that music, art and humans have, from their earliest beginnings, been completely entangled is compelling. What is more, the sense that these entanglements have endured, twisting and continuing to evolve over millennia, reminds us that – as with Barad's investigations into quantum measurement – it is the boundaries that are formed by humanity's embrace of music that we must remain mindful of, rather than being lured into focusing on essential qualities. These boundaries change, and as they change so too does our relationship with music. Indeed, in the context of such millennia-long entanglements we can also consider our own entangled relationship with our planetary habitat, and with ourselves.

Humans after the Anthropocene

As we saw in the opening pages of this book, in 2019, the Anthropocene Working Group (AWG), confirmed the geological reality of the Anthropocene, establishing 1950 as its approximate threshold. The AWG cited the post-war 'Great Acceleration' which facilitated the availability of cheap oil, polymer-chain products and radioactive fallout from nuclear weapons tests as the key stratigraphic markers for the beginning of this new period (Anthropocene Curriculum, 2022). While the AWG has for the time being settled the debate as to the Anthropocene's point of origin, a variety of alternatives have been offered, which reflect the dynamic, and often volatile, relationship that we humans have had with the planet earth throughout our existence.

In 2000, the atmospheric scientist Paul Crutzen and the biologist Eugene Stoermer determined that 1784 marked the onset of the Anthropocene era, aligning the first traces of humanity's ability to alter the balance of gases in earth's atmosphere – specifically increases in the greenhouse gases carbon dioxide (CO2) and methane (CH4) – with James Watt's invention of the steam engine (Crutzen and Stoermer, 2000). The environmental activist George Monbiot presents an alternative narrative for the emergence of measurable human impact on the planet, based on analyses of pollen core samples extracted from the Clwydian Hills in North East Wales. In the book *Feral* (2013) he describes how the core samples taken in 2007 provide evidence of the widening spread of hazel, oak, alder, willow, pine and birch in the hills, after the retreat of the last ice age, 8,000 years ago. Initially, the trees accounted for 30 per cent of the pollen found in the area, rising to 70 per cent 4,500 years ago. As Neolithic farmers began to make use of the land, gradually the tree pollen began to decline to be replaced by heather pollen, to the extent that, by 1900 heather accounted for 60 per cent of the core, compared with 10 per cent tree pollen. For Monbiot, this demonstrates that 'the open landscapes of upland Britain, the heaths and moors and blanket bogs, the rough grassland and bare rock which many people see as the natural state of the hills, are the result of human activity' (Monbiot, 2013: 66–7). In *The Shock of the Anthropocene* (2017) Christophe Bonneuil and Jean-Baptiste Fressoz also recognise the late 1700s as a boundary point that marks the onset of the 'carbonification' of our atmosphere, but go on to reflect on other possible origins of the Anthropocene, including the advent of Homo Sapiens 200,000 years ago, and a spike in greenhouse gas emissions caused by deforestation, livestock rearing and crop production roughly 5,000 years ago (Bonneuil and Fressoz, 2017).

Beyond these possible alternatives to marking the onset of the Anthropocene, Bonneuil and Fressoz, along with environmental historian Jason Moore, have aimed to unbalance the idea that humanity as a uniform and homogeneous *anthropos* is the cause of the Anthropocene. Moore uses the concept of the 'Capitalocene' to fracture such an anthropos; seeing the origins of the Anthropocene in the rise of capitalist civilisation after 1450, rather than in Watt's steam engine and coal extraction (Moore, 2015: 172). In essence, Moore sees that the power relations of 'fossil capitalism' are the true cause of climate change: 'shut down a coal plant, and you can slow global warming for a day; shut down the relations that made the coal plant, and you can stop it for good' (ibid.). Bonneuil and Fressoz partition the anthropos in a different way, using the notion of an 'environmentalism of the poor' to fragment the idea of a uniform humanity, all pulling together as a homogeneous industrial-capitalist whole towards a warmer planet (Bonneuil and Fressoz, 2017: 253). They contrast the idea of 'private landowners and royal foresters' fixed on rationalising forestry management for profit (ibid.: 255) with the 'machine-breaking movement' of groups such as the Luddites and the Chartists (ibid.: 259) to assert that, historically, the global poor were not invested in an anthropogenic technologically driven industrialism in the same way as the capitalist elite.

Whether we see the beginnings of the Anthropocene in the post-war profusion of oil, plastics and nuclear waste, in the development of James Watt's steam engine, in the emergence of Capitalism in the late Middle Ages, or in the decimation of pre-human habitats brought on by Neolithic farmers introducing ruminants from the Fertile Crescent to defenceless European habitats, my intention here is to bring a focus to how human practices of disrupting, modifying and shaping our environment at local, global and atmospheric scales are activities that we have engaged with for millennia. As we see with Sheldrake's and Lowenhaupt-Tsing's investigations into the interdependent nature of the living world and of our socio-economic systems and cultures, it appears that humans cannot avoid modifying the world, and at the same time, we are unavoidably modified by the world.

I began *Ecologies of Creative Music Practice* by suggesting that the Anthropocene, beyond signifying a paradigm shift in humanity's capacity to fundamentally alter the earth's ecosystems, also marks a change in how we understand ourselves in relation to concepts of 'nature', and our own place in a planetary ecosystem of nonhuman life. Our ability to understand and conceptualise the Anthropocene demonstrates our capacity to not only acknowledge the impact that human development has had on our atmosphere, and therefore on our planetary ecosystems, but also to recalibrate our understanding of our relation as humans to our planetary environment, and consider what this means in terms of our understanding of ourselves. In the first instance, Crutzen and Stoermer's pathfinding work and initial formulation of the concept of the Anthropocene enabled us to confront the exponential rate of climate change catalysed by human industrial activity. Second, in a more subtle, but no less important sense, the existence of the Anthropocene as a concept both marks and makes a paradigm shift in our understanding of ourselves. Rather than seeing nature as a resource that humans simply interact with, Moore frames it as a matrix through which humans develop, such that 'relations between humans are themselves produced through nature' (Moore, 2015: 124). This reflects Barad's proposal that intra-action is the process by which an objectively real world is produced, which is to say that we are part of nature, and yet it is a real environment that exists independently of our capacity to think or engage with it. Our entanglement with nature produces our ability to understand ourselves as both part of and separate from it.

In the essay 'Some Trace Effects of the Post-Anthropocene' (2013), Benjamin Bratton goes even further, suggesting that is that it is not humanity as an undivided whole that the Anthropocene destabilises – which we see in Moore's and in Bonneuil and Fressoz's fragmentations of a unified anthropos – but the very idea of humanity as a self-same concept. In this sense, the Anthropocene does not simply mark a radical change in the environment. It also marks our recognition of our ability to impact on the real development of the planet, a shift in awareness which means that humanity itself is no longer what it was. Where Moore sees that the Anthropocene has led us to understand that human cognition is part of a nature matrix, Bratton sees that our part in this matrix – what he describes as an expanded conception of the human phylogenome - will lead to 'a more open-ended convolution toward adaptation, invention, diversion, and reiteration [and that] should "we" survive the Anthropocene, it will not be as "humans"' (Bratton, 2013: 5). Beyond its use as a term to describe the impact that humans have had on earth systems, the concept of the Anthropocene therefore becomes a statement about what constitutes the idea of the human, as much as it is about humanity's impact on the planet. For Bratton, in creating the Anthropocene, we have created a new type of human; we are no longer the humans we once were.

From these perspectives emerges the sense in which the Anthropocene itself is a boundary and a mark imprinted on humans, that has been produced by the collision of our planetary environment with human-engineered forces and systems of organisation. As we saw with Reza Negarestani's theories concerning the artificial nature of intelligence, such a capacity for self-awareness is the defining feature of artificial general intelligence (AGI), a self-awareness that recognises self-identity, and the artificial nature of that identity. To be intelligent is to recognise oneself as being separate from the world, and at the same time understand that one's intelligence is formed by the collective dynamics of particular and located forces. This is to say that, just as humans have produced the Anthropocene, the Anthropocene has produced a certain type of intelligence. In the context of music, such a reciprocity can be discerned in the way that Spitzer sees that our relationship with sound and music has produced a certain type of intelligence, and how, throughout time, humans have continually reformulated what music is, and what it can be; bone flutes in a cave, the sounds of our own nervous system in an anechoic chamber, a 25 kw bassline moving up our legs, or the ambient sounds in an abandoned nuclear reactor reflecting back on themselves.

Ray Brassier and Reza Negarestani extend the challenge of articulating human/nature relations that the Anthropocene brings into view, destabilising the idea of a uniform anthropos in a different way via a process 'un-humaning' humans. In the short essay 'Prometheanism and Its Critics' (2014), Brassier outlines what he calls a 'Promethean project', informing us that, 'Prometheanism is simply the claim that there is no reason to assume a predetermined limit to what we can achieve, or to the ways in which we can transform ourselves and our world' (Brassier, 2014: 470). The contention is that, in the same way that the Anthropocene has demonstrated that humans can shape the ecology of our environment, it is also possible for us to shape ourselves. In essence, Brassier's Prometheanism is an insistence that the capacity to think beyond, and transform ourselves is intrinsic to our nature. That we are now a people of the Anthropocene is a demonstration of our ability to evolve ourselves. For Brassier,

> Prometheanism is the attempt to participate in the creation of the world without having to defer to a divine blueprint ... the disequilibrium we introduce into the world through our desire to know is no more or less objectionable than the disequilibrium that is already there in the world.
>
> *(ibid.: 485)*

In other words, Prometheanism is a way of understanding that our embeddedness in the world means that we are part of its inherent disequilibrium, and that this disequilibrium is inherent in us. Simply put, Prometheanism means that making the world – making music, making artificial intelligence, making blockchain tokens and making the Anthropocene – is the process through which human thought transforms itself into something other than it was.

In the essay 'The Labor of the Inhuman' (2014), Negarestani maps out a concept that he refers to as 'inhumanism', an idea designed to show that, as we have seen in Meillassoux's work, humans are capable of producing thought that is without limits. In other words, we are able to conceive of the existence of that which is not a product of human thought and which is beyond our ability to comprehend. What is more, because of this capacity, Negarestani proposes that human thought is a necessary agent of real change in the world. The inhuman is a way of framing human reason and rational thought as the 'discontinuous [...] content of humanity' (Negarestani, 2014: 450). This means that, in order to fully appreciate how human thought operates, we should understand that rational thought is an 'autonomous space'; an impulse that creates 'catastrophes' and 'ruptures' that form new limits and boundaries in our understanding of the world (ibid.). Such an autonomous space implies a kind of necessary exponentiality, something that sits outside of our control; a version of human rationality and reason that is not predictable. This seems counter-intuitive. We imagine ourselves to be in control of our own reason, since, surely, reason is fundamentally a controlling mechanism. However, Negarestani sets reason up as an inhuman, impersonal force that acts on the human, such that 'inhumanism [becomes] the labor of rational agency on the human' (ibid.: 446). As we have seen with Negarestani's views on the artificiality of intelligence, his concept of inhumanism also speaks of the unfinished nature of the human entity of human thought as a work in progress: 'the kernel of inhumanism is a commitment to humanity via the concurrent construction and revision of the human as oriented and regulated by the autonomy of reason' (ibid.). What is interesting here is that, by creating this divide in how thought *thinks*, Negarestani is positioning the capacity of human thought to think both outside of itself and in on itself, while always remaining completely itself, as the defining feature of humans that enables us to overcome the problem of using thought to engage with something that is beyond thought. In essence, Negarestani is contending that the human 'is not itself'; that inherent in us is a quality that contradicts conventional notions of what the human is, which also suggests a further flaw in the idea that the anthropos is a unified articulation of humanity, more fundamental than Jason Moore's focus on the pernicious influence of Capital. Before we arrive at the anthropogenic tendencies of a capitalist system, the anthropos is already always contingent; fragmented by inhuman, autonomous reason and the discontinuous nature of rational thought.

Where Negarestani invokes a self-radicalising insider thought that uses the 'in' in the inhuman, to outline an alternate vision of humanism, Brassier tells us that, in essence, Prometheanism is a recognition that humanity's capacity to disrupt and alter the world reflects the world's inherent disequilibrium, a mirroring that is not unlike the Laruellean concepts of cloning and fictioning, wherein humans perform the Real through creative acts in the real world of everyday experience. These perspectives again directly connect to our earlier discussions of the causes and consequences of the Anthropocene, where, given that a tendency towards disequilibrium is a determining and inherent feature of humanity, then a discontinuous, rather than undifferentiated, anthropos is both a catalyst for and a result of the nature–human–Anthropocene matrix. In other words, we have always been creating discontinuous effects in both ourselves and our environment, and the series of anthropogenic events that Bonneuil and Fressoz detail are simply part of an even longer history of environmental disruption that humans are both the result and the cause of. As creatures that exists in a contingent world, and who are grounded by the contingency of that world, then we humans are living systems in motion, playing out that fundamental contingency through our thought and behaviours.

Brassier's and Negarestani's ideas lead us down a demanding path, directing us to engage with inherent contingency and discontinuity as essential qualities in the world. As part of that world, we humans cannot help but be shaped by and create the world according to its discontinuous nature, a nature that exists beyond our thought, but at the same time is the foundation of our thought. Whether we think in terms of Promethean disequilibrium or inhuman discontinuity, we can think of contingency as the instability and indeterminacy of the world folding back through us, an idea that reconnects us with Meillassoux's great outdoors, Laruelle's One, and Barad's quantum indeterminacy. However we frame the inescapably contingent or indeterminate nature of the world, the arrival of the Anthropocene marks a pivotal point in our ability to understand the world, and our place within it. And as with the Anthropocene, so with music. Both are the results of human making practices that, because of our discontinuous and contingent natures, are practices of creating discontinuity, contingency and disequilibrium in the world. As we have seen throughout the chapters in *Ecologies of Creative Music Practice*, just as the Anthropocene, music, as a contingent product of human disequilibrium, can do nothing else but force change and adaptation, in order that we humans can begin to comprehend what we have made.

Quantum Ecologies of Creative Music Practice

Throughout this book, I have explored how a variety of creative practices of music are shifting and evolving in response to a range of factors that mark music out as an ecology in motion. However, thinking ecologically does not mean simply understanding that music is connected to a vast range of creative, technological, economic and social practices. Instead, it means that music comes into being in new ways every time its relations with the other forces and nodes in its ecosystems evolve and change. In the context of our examination of quantum theory and materialist philosophies, since dynamic flows that comprise ecosystems are material forces, I will conclude by extending the analysis to consider music as a quantum ecology.

As we have seen in Karen Barad's work, quantum theory is fundamentally a theory of measurement, and interactions between things can therefore be understood as practices of observation and measurement. At the same time, such practices are productive. Measuring and observation, as practices of interaction, produce boundaries. Measurements leave marks on bodies. Thus, as an ecology, music is the measurement of interactions, and interactions produce boundaries; they leave marks. Music is therefore both produced and productive of myriad boundary-producing interactions. For example, the streaming platform Spotify is not simply the result of an interaction between music, technology and the digital landscape of contemporary commerce. Instead, Spotify has resulted from the constituent parts of music, technology, economics; interacting and producing a certain measurement of music that then produced Spotify. To think in terms of quantum ecologies of creative music practice is therefore to understand that music is not made up of things that are interacting; it is made up of interactions and measurements.

Carlo Rovelli describes the development of our understanding of reality as a way of thinking about 'a reality that is more subtle than the simplistic materialism of particles in space. A reality made up of relations rather than objects' (Rovelli, 2021: 3). In this sense, practices of music *are* the relations between music, technologies, economics, cultures, materialist and speculative realist discourses, and natural environments and ecosystems. For Rovelli,

> The properties of an object are the way in which it acts upon other objects; reality is this web of interactions. Instead of seeing the physical world as a collection of objects with definite properties, quantum theory invites us to see the physical world as a set of relations. Objects are its nodes.
>
> *(ibid.: 70)*

Rovelli's nodes are Barad's points of measurement; moments formed through creative practices of agential realism. They are the creator's practice of making. They are the listener's listening. They are the point of contact between AI and music production, between TikTok and practices of production, communication, remixing, earning a living. Thus, TikTok is a node of agential realism, as is LANDR, Musicoin, ROCKI, *Recordings from the Aland Islands, Universal Beings, A Norfolk Rhapsody*, a reconstruction of an ancient Welsh Harp, Spotify, Resonate and Hipgnosis Songs Fund. As with Brassier's and Negarestani's claims about the inherent disequilibrium and discontinuities that structure both the world and human thought, Rovelli too sees that 'the world is not continuous, but granular' (ibid.: 96). His and Barad's quantum theorisations enable us to understand that the world is not discontinuous because it is made up of discrete objects. It is discontinuous because the interactions between objects (nodes) are discontinuous. Measurement, observation and creative mark-making are not continuous; these operations only occur at the point of intra-action between objects.

The wager in *Ecologies of Creative Music Practice* is that the world is not simply reducible to a set of natural components, and that, therefore, humans, artificial intelligence, Vaughan Williams' *Norfolk Rhapsody*, Mark Guiliana's acoustic drumkit, Bob Dylan's song catalogue, or Japanese cover versions of Justin Bieber songs on TikTok are all nodes, objects, boundaries and marks on bodies in a world fundamentally composed of relations.

However, before we get too carried away, as Robin James has shown, we must take care when using new materialism as a thinking tool. If everything is essentially composed of the same material forces, it can be challenging to establish a framework for building either an objective perspective or simply allowing for localised differentiation. In short, if everything is fundamentally the same, and technology, market forces, environmental conditions, human actions are all intra-actively entangled, then how can anything ever be different, and how do we create anything that is objectively and qualitatively different from a homogenised mass of material entanglements? Lovelock and Negarestani can help us here. For each of them, intelligence is a matter of a certain type of exceptionalism. As we have seen, while intelligence is a defining quality of humans, this does not mean that human intelligence is the only possible form of intelligence. In a sense, intelligence itself is a unique property, and to a degree we could think of humans as a 'carrier wave' of intelligence. However, intelligence is not simply an object that exists for-itself. In the same way that intelligence is a marker of a non-human reality – again, what Rovelli refers to as a 'node' – it is the human understanding of intelligence, and – as Negarestani suggests – the discontinuity of rational thought and its destabilisation of familiar conceptions of the human that demonstrate the intelligence itself is a process under construction.

Drawing on the contributions of the physicist Ernst Mach in the context of early twentieth-century traditions of materialism, Rovelli examines Mach's view that 'sensation' is real, rather than a material world of 'matter in motion' (ibid.: 107). Essentially Mach's idea was that the human subject is formed in response to sensation; in other words, the subject comes into being as a result of an objectively existing interaction – what Barad would call an intra-action – with something other than itself. This is qualitatively different in kind from the idea that the self is a subjective 'I' that experiences the world directly, and is the only qualifier of that world (in other words, correlationism). If measurements, interactions or Mach's 'sensations' leave marks on bodies, then superposition is the state of not-being-measured. Not to be measured means having no interactions with anything else. As Rovelli suggests, 'to speak of objects that never interact is to speak of something that could not concern us' (ibid.: 68). In this sense it is difficult to say whether objects that do not interact with anything else can be said to exist, or whether objects which do not interact with other objects are property-less. Following Rovelli, we can understand that superposition does not mean that objects adopt a different form of interaction with the world that humans cannot observe or understand, for example, a wave. Instead, superposition means that objects stop interacting with anything. They are as good as 'not being there'.

A piece of music that features coastal sounds from the Åland Islands, a recording of musicians improvising in a church, all edited together to achieve a certain sonic outcome is the subscendant digital sum of multiple interactions and entangled measurements. A blockchain music streaming platform that allows creators to pay their listeners, the growing popularity of legacy music over new music, AI-powered creative music apps that enable people with no musical ability but with money to spend on hitting a deadline, and social media platforms that facilitate and monetise likes, reactions, reposts – all of these are quantum ecologies. Ecosystems are therefore entangled measurement systems. The forces engage and produce tangible boundaries at the point of measurement. Ultimately, without this cascade of measurement-interactions, music itself would not exist. It is simply the ongoing result of interactions, boundary-making practices, observations, creative engagements. Nothing about music is fixed, and nothing about it has ever been fixed. Music is inherently in motion, and if it were not for this motion, then – as Rovelli suggests – it may well cease to exist.

This brings us to the idea that 'sensation leaves marks on bodies'; in this case, the self. How the self responds to those marks is another matter. But the fundamental point here is that the world – in the form of sensations – makes the subject; the individual is not creating subjective versions of an ideal world. The self can therefore be seen as an ecology of sensations, rather than a point of view that constructs the world according to its own perspectival capacities and limitations. This intuition for the 'impersonal' being that which we all have in common, is similar to how Rovelli reads Mach's ideas about the material, objective reality of sensations, as opposed to matter, and individual things as such. Mach's sensations are Barad's measurements and observations, and agential realism is the experience of this non-subjective, real exterior to thought, a moment of engagement between an objective real and subjective experience that produces – via what Barad calls 'boundary making practices', the kinds of things we call 'music', 'volume', 'speed', time signatures', 'field recordings', 'Spotify', 'song catalogues', 'F# major', 'distortion', and categories such as 'music creators', 'listeners', 'musicians' along with the fundamental concept of 'the individual'. In this sense, reality comprises subjectively experienced, objectively existing relations which produce objectively existing outcomes.

To say that 'music isn't music' would be problematic. However, it speaks to an awareness that the mattering of music via a quantum ecological analysis results in an acknowledgement of both music's interdependence with a vast range of forces, along with the realisation that music itself is unfixed and indeterminate. Making music is not the practice of producing a new piece of music – a performance, a song, an NFT, a digital file, an Instagram Reel, a TikTok video – rather, it is the practice of producing the possibility of music. Music comes into being every time we make it. A quantum ecological history of music is thus not simply the history of different styles of music being made, it is the history of the possibility of what music is, expressed via the production of every musician's and every listener's vision for what it can be. We don't know what music is, but given the extent to which social, economic and creative technologies have exponentially opened up opportunities for creation, listening, watching and sharing, we can be confident that the indeterminacy and interdependence of music will continue to provide rich contexts for practice and debate.

Notes

1 Sheldrake provides several examples of fungi altering the behaviour of animals, including the way in which *Ophiocordyceps unilateralis*, which he refers to as 'zombie fungus', is able to manipulate the movement of carpenter ants. The fungus infects an ant, and while living inside it, forces it to climb up the nearest plant stem; something the ant would not normally do. The fungus forces the ant to clamp its jaws around the stem, before growing through the ant's feet and into the stem. Finally, the fungus digests the ant's body, and grows a stalk out through the ant's head, from which it is able to drop spores onto any ants that are underneath, thus starting the cycle again (Sheldrake, 2020: 107).

2 Derek Bailey is recognised as one of the key figures in post-war British experimental and improvised music, and his concept of 'non- idiomatic' improvisation made a significant contribution to the language and discourse surrounding improvised music practice. Indeed, his naming of a certain approach to music making as non-idiomatic has shaped how musicians engage with, understand and create music spontaneously, outside of more obviously context-based improvisational practices, such as jazz, Indian classical music, flamenco and, to an extent, rock music.

3 The anthropologist David Graeber and archaeologist David Wengrow support this position in their suggestion that it is only after archaic human populations such as Neanderthals and Denisovans had completely died out, by approximately 40,000 BCE, that we can talk with any authority about modern humans as the only species of human inhabiting the planet (Graeber and Wengrow, 2021). Only at this point in time can we begin to make credible statements about a self-same species of human that exists up to the present day.

Bibliography

Anthropocene Curriculum. 2022. 'Anthropogenic markers'. Available at https://www.anthropocene-curriculum.org/anthropogenic-markers (accessed November 2022).

Barabási, A-L. 2016. *Network Science*. Cambridge: Cambridge University Press. Available at http://networksciencebook.com (accessed November 2022).

Bonneuil, C. and Fressoz, J-B. 2017. *The Shock of the Anthropocene: The Earth, History and Us*. London, New York: Verso.

Brassier, R. 2014. 'Prometheanism and its critics', in Mackay, R. and Avanessian, A. (eds), *#Accelerate: The Accelerationist Reader*. Falmouth: Urbanomic.

Bratton, B. 2013. 'Some trace effects of the post-Anthropocene: On accelerationist geopolitical aesthetics'. Available at https://www.e-flux.com/journal/46/60076/some-trace-effects-of-the-post-anthropocene-on-accelerationist-geopolitical-aesthetics/ (accessed November 2022).

Crutzen, P.J. and Stoermer, E.F. 2000. 'The "Anthropocene"'. *Global Change Newsletter*, 41: pp. 17–18.

Graeber, D. and Wengrow, D. 2021. *The Dawn of Everything: A New History of Humanity*. London: Allen Lane.

Lowenhaupt Tsing, A. 2015. *The Mushroom at the End of the World: On the Possibility of Life in Capitalist Ruins*. Princeton, NJ: Princeton University Press.

Marley, B. and Wastell, M. 2005. *Blocks of Consciousness and the Unbroken Continuum*. London: Sound 323.

Meadows, D. 2008. *Thinking In Systems: A Primer*. White River Junction, VT: Chelsea Green Publishing.

Monbiot, G. 2013. *Feral: Searching for Enchantment on the Frontiers of Rewilding*. London: Penguin.

Moore, J. 2015. *Capitalism in the Web of Life: Ecology and the Accumulation of Capital*. London, New York: Verso.

Negarestani, R. 2014. 'The labor of the inhuman', in Mackay, R. and Avanessian, A. (eds), *Accelerate: The Accelerationist Reader*. Falmouth: Urbanomic.

Rovelli, C. 2021. *Helgoland: The Strange and Beautiful Story of Quantum Physics*. London: Allen Lane.

Sheldrake, M. 2020. *Entangled Life: How Fungi Make Our Worlds, Change Our Minds and Shape Our Futures*. London: The Bodley Head.

Spitzer, M. 2021. *The Musical Human: A History of Life On Earth*. London: Bloomsbury.

INDEX